Nanobiotechnology for Safe Bioactive Nanobiomaterials

This book begins with an introduction of nanobiotechnology, followed by bio-syntheses of AgNPs, development of silver/chitosan (Ag/CS) polymer nano-composites, synthesis of silver/chitosan-g-poly acrylamide (Ag/CS-g-PAAm) nanocomposite hydrogel and silver/chitosan/poly vinyl chloride (Ag/CS/PVC) blend. Finally, it presents novel bioengineering of polyfunctional metallic nano-structures other than Ag, emphasizing biomass utilization and value-added con-version over an extended span, including life cycle assessment of the synthesized nanostructures.

Features:

- Includes prospective cost-effective, eco-friendly and safe nanomaterials, synthesized through facile paths
- Covers the synergistic effect of phytochemicals and nano-Ag antimicro-bial agents from an antiviral perspective
- Includes surface coating systems and super absorbent materials for bio-medical purposes
- Examines nanobiotechnological applications for generating nanoalloys with synchronized nanostructural arrangement of alkaline earth metals and nanoscale dots of transition metals
- Explores the life cycle assessment of synthesized nanomaterials

This book aims at researchers and graduate students in biomaterials, chemical engineering, green chemistry, nanomaterials and biotechnology.

Novel Biotechnological Applications for Waste to Value Conversion

Series Description:

Solid waste and its sustainable management is considered as one of the major global issues due to industrialization and economic growth. Effective solid waste management (SWM) is a major challenge in the areas with high population density, and despite significant development in social, economic and environmental areas, SWM systems are still increasing the environmental pollution day by day. Thus, there is an urgent need to attend to this issue for green and sustainable environment. And, the proposed book series is a sustainable attempt to cover waste management and the conversion of waste into value-added products.

Series Editors:

Neha Srivastava
IIT BHU Varanasi, Uttar Pradesh, India

Manish Srivastava
IIT BHU Varanasi, Uttar Pradesh, India
Utilization of Waste Biomass in Energy, Environment and Catalysis

Dan Bahadur Pal and Pardeep Singh
Nanobiotechnology for Safe Bioactive Nanobiomaterials

Poushpi Dwivedi
Department of Chemistry, Belda College, Vidyasagar University, West Bengal, India

Shahid S. Narvi
Department of Chemistry, Motilal Nehru National Institute of Technology Allahabad, Uttar Pradesh, India

Ravi Prakash Tewari
Department of Applied Mechanics, Motilal Nehru National Institute of Technology Allahabad, Uttar Pradesh, India

Dhanesh Tiwary
Department of Chemistry, Indian Institute of Technology (Banaras Hindu University), Uttar Pradesh, India

For more information about this series, please visit: https://www.routledge.com/Novel-Biotechnological-Applications-for-Waste-to-Value-Conversion/book-series/NVAWVC

Nanobiotechnology for Safe Bioactive Nanobiomaterials

Poushpi Dwivedi, Shahid S. Narvi,
Ravi Prakash Tewari and Dhanesh Tiwary

CRC Press
Taylor & Francis Group
Boca Raton London New York

CRC Press is an imprint of the
Taylor & Francis Group, an **informa** business

First edition published 2023
by CRC Press
6000 Broken Sound Parkway NW, Suite 300, Boca Raton, FL 33487-2742

and by CRC Press
4 Park Square, Milton Park, Abingdon, Oxon, OX14 4RN

CRC Press is an imprint of Taylor & Francis Group, LLC

ISBN: 9781032108452 (hbk)
ISBN: 9781032108469 (pbk)
ISBN: 9781003217343 (ebk)

DOI: 10.1201/9781003217343

Typeset in Times
by KnowledgeWorks Global Ltd.

Dedicated

to

Shri Kashi Vishwanath

Contents

       5.3.1   FTIR Elucidation...112
               5.3.1.1   FTIR Study of the Silver
                         Nanoparticles 112
               5.3.1.2   FTIR Study of the Ag/CS-g-PAAm
                         Nanocomposite 114
       5.3.2   Characteristic Study of the Ag/CS-g-PAAm
               Nanocomposite Hydrogel Super Absorbent
               Polymeric (SAP) Material 114
       5.3.3   Mechanical Property 116
       5.3.4   Swelling Behavior 116
       5.3.5   Quantitative Evaluation of Ag$^+$ Release 116
       5.3.6   Antimicrobial Assay..117
       5.3.7   In vitro Blood Compatibility 117
5.4   Principle of Application...118
5.5   Conclusion 118
References 119

Chapter 6   Bioactive Silver/Chitosan/Polyvinyl Chloride (Ag/CS/PVC)
            Nanocomposite Blend: Phytomass Enabled 121

       6.1   Introduction 121
       6.2   Experimental 123
              6.2.1   Materials..123
              6.2.2   Synthesis – Synthesis of Ag/CS/PVC
                      Nanocomposite Blend...123
              6.2.3   FTIR Spectroscopy 124
                      6.2.3.1   FTIR Study of the Silver
                                Nanoparticles 124
                      6.2.3.2   FTIR Study of the Ag/CS/PVC
                                Nanocomposite 124
              6.2.4   Characterization of Ag/CS/PVC
                      Nanocomposite..124
              6.2.5   Swelling Parameters...124
              6.2.6   Quantitative Evaluation of Ag$^+$ Release 125
              6.2.7   Mechanical Testing 125
              6.2.8   Antimicrobial Assay...125
              6.2.9   In vitro Blood Compatibility Test.......................125
       6.3   Results and Discussion 126
              6.3.1   FTIR Elucidation..126
                      6.3.1.1   FTIR Study of the Silver
                                Nanoparticles 126
                      6.3.1.2   FTIR Study of the Ag/CS/PVC
                                Nanocomposite 127

                     Nanocomposite 129
         6.3.3   Swelling Behavior 130
         6.3.4   Quantitative Evaluation of Ag⁺ Release 131
         6.3.5   Mechanical Properties..131
         6.3.6   Antimicrobial Assay..131
         6.3.7   *In vitro* Blood Compatibility 132
6.4   Principle of Application... 133
6.5   Conclusion 133
References 133

Chapter 7   Prospects of Safe Functional Nanomaterials in the Era
                 of Virus Dominion 137

         7.1   Overview 137
                7.1.1   Synthesis – Phytomass Conversion and
                         Value-Added Fabrication of Functional
                         Nanomaterials 141
                         7.1.1.1   Materials 141
                         7.1.1.2   Synthesis of Ag/CS
                                    Bionanocomposite Coating Material.......141
                         7.1.1.3   Coating of Object Surfaces.................. 142
                         7.1.1.4   UV-Visible Spectrometric Study 142
                         7.1.1.5   HR TEM and SAED Observations
                                    with EDX Analysis 142
                         7.1.1.6   Elucidation of Physico-Chemical
                                    Properties through FTIR, SEM
                                    and XRD... 142
                         7.1.1.7   Antimicrobial Assay 143
                         7.1.1.8   Cytotoxicity Test and
                                    Biocompatibility Assessment 143
                7.1.2   Characterization 143
                         7.1.2.1   UV-Visible Spectral Analysis 143
                         7.1.2.2   Qualitative Assessment by HR
                                    TEM, SAED and EDX 145
                         7.1.2.3   Study of Physico-Chemical
                                    Parameters through FTIR, SEM
                                    and XRD... 145
                         7.1.2.4   Antimicrobial/Biological Activity....... 146
                         7.1.2.5   Biocompatibility Assessment 147
                7.1.3   Antiviral Hypothesis 148
         7.2   Principle of Application... 148
         7.3   Future Scope – Focusing Antiviral Perspective
                during the Virus Dominion 149
         References 150

Preface

The confluence of nanotechnology with material science, biology, biotechnology and medicine has paved the path to explore the dark avenues and solve the toughest problem in almost every field including biomedical engineering, thereby improving the 'quality of life', globally. Today, the most alarming global problem in the biomedical arena is bacterial infection at the site of implanted medical devices, prosthetics and sensors. Despite aseptic measures and sterilization procedures, microbial infection poses a major impediment to the utility of the biomaterials.

Therefore, in this nanoregime and virus era, present research has endeavored to bring forth antimicrobial, self-sterilizing biocompatible nanomaterials via nanobiotechnology, chiefly utilizing the wealth of phytomass and value-added conversion of other renewable biomass, which can indubitably be used for significant applications, especially for biomedical as safe bioactive nanobiomaterials. The developed nanobiomaterials are cost-effective and eco-friendly, produced through facile path, avoiding the use of obnoxious reagents, instead include the biomass beneficiaries, while following the principles (1, 3, 4, 5, 6, 7, 8, 10 and 12) of 'Green Chemistry'. Thus, an original winning strategy has been developed to mainly combat the prevailing race between bacterial adhesion and tissue integration to potentially mitigate the menace of biomaterials associated infection (BAI). Hypothetical postulation has also been endorsed to battle out popping mutant viruses through surface modification, restricting substantial contamination and transmission, replacing repeated sanitization procedures.

CHARACTERIZATION TECHNIQUES

Various techniques have been used for characterization, such as UV-visible spectroscopy, energy dispersive X-ray (EDX) analysis, transmission electron microscopy (TEM), nanoparticle size analysis, scanning electron microscopy (SEM), X-ray diffraction (XRD), Fourier transformed infrared (FTIR) spectroscopy, differential scanning calorimetry (DSC), thermo-gravimetric/differential thermal analysis (TG/DTA), mechanical test, scratch test, swelling ratio, atomic absorption spectroscopy (AAS), antimicrobial assay, cytotoxicity test, blood compatibility test, etc. Evaluations through these parameters are essential factors for establishing the structure and the biocompatibility of the synthesized materials.

ORGANIZATION OF THE BOOK

The monograph has been presented in nine chapters. Chapter 1 is of general introduction, dealing with the idea of nanotechnology, nanomedicine and nanobiotechnology. This chapter also describes nanomaterials, particularly polymer nanocomposites and nanobiomaterials, along with the required techniques for characterization and their scope of biomedical applications, highlighting the importance of silver, silver nanoparticles and silver nanocomposites together with their assessment of risk factors and biocompatibility. Chapter 2 consists of the concise literature review, literature gap and research objectives. Chapter 3 deals with the work done for the nanobiotechnological syntheses of silver nanoparticles, through utilization of phytomass for bio-reduction of silver ions (Ag^+) to silver atoms (Ag^0) with diverse phytomass extracts followed by colloidal aggregation and self-assembly for the generation of nanoparticles (phytofabrication), purposing a comparative study. Chapter 4 consists of the work carried out to develop the silver/chitosan (Ag/CS) polymer nanocomposite via nanoiotechnological route using foliage needles of the plant *Pseudotsuga menziesii*, for coating of biomedical implants, along with the study of its efficacy against *Staphylococcus aureus* biofilm. Chapter 5 deals with the synthesis of silver/chitosan-g-poly acrylamide (Ag/CS-g-PAAm) nanocomposite hydrogel through graft copolymerization technique and utilizing rhizome phytomass of the plant *Curcuma longa*. Chapter 6 contains the work done for the synthesis of nanocomposite blend, silver/chitosan/poly vinyl chloride (Ag/CS/PVC), utilizing the bark of *Terminalia arjuna*, demonstrating overwhelming antibacterial response. Chapter 7 discusses future prospects of the present studies, along with utilization of the foliage of *Ocimum tenuiflorum* (holy basil/Tulsi) for value enhancement of the coating system synthesized with antiviral efficiency and hypothesis in the context of virus dominion. Chapter 8 chiefly emphasizes the valorization of human finger and toe nail pruning biomass (HNB) into magnificent nanostructures of significant MgO/CaO nanoalloy. Last but not the least, Chapter 9 concludes the monograph, speculating the environmental fate of the safe nanobiomaterials, synthesized through nanobiotechnology, portraying the cradle to grave, life cycle assessment (LCA) of the bioengineered nanostructures.

SILVER LINING IN THE ERA OF EMERGING NOVEL VIRUSES

The phytochemicals aided, nanobiotechnologized cost-efficient, eco-friendly, self-sterilizing green nanomaterials accounted, enable prolonged discharge of the silver ion (Ag^+) biocide at concentration level of <0.1 ppm, capable of imparting long-term enhanced antimicrobial efficacy and hardly a few microbes can be intrinsically resistant. Our innovation of the functional nanomaterials, such as the self-sterilizing surface coating system, other nanocomposite biomaterials and nanostructured alloys have industrial scale-up feasibility. They are the much in demand products globally, indicating relieve measures, casting respite to the expensive and tedious regular process of repeated sanitization and sterilization. Therefore, the greatly required value-added conversion employing nanobiotechnological pathway is concisely presented here, from renewable bioresources to bioactive nanobiomaterials consisting antiviral prospects.

Acknowledgments

The authors believe that God's grace has enabled to undergo the research, unveil the unsearched and compile this work. Research is a passion while service to humankind a mission. Poushpi Dwivedi (PD) desperately aspired to coordinate the twin desire, and in an attempt to do so, she was deeply inspired by the 'next big small thing' termed as 'nanotechnology'. In addition, she felt the urgent need to endorse 'green chemistry' and consequently proceeded toward nanobiotechnology for the green and safe bioactive nanobiomaterials. Also, PD realized research is definitely a teamwork which can't be achieved by a one man army. The Almighty had only coordinated all noble and able people who helped to find the novel and to write this book.

The first author, PD, is grateful to her parents, especially her father, Prof. Jayan Shankar Dutta, who is like a pillar and for the countless things he did, who gave the world to her, but tears roll down her eyes as she wishes to express her immense gratitude to the Late Er. Brijesh Dhar Dwivedi, the father she got after marriage, who made the world for her. PD also expresses her thanks to her husband, Dr. Muktesh Dwivedi, who, behind the curtain, in some or the other form, made this mission possible, and words do not suffice to render PD's thankfulness for her little children, Akkshita Dwivedi and Amritesh Dwivedi, as they have huge contribution in their nano ways.

PD earnestly expresses her deep sense of gratitude to Dr. Manabendra Mondal, Principal, Belda College, Belda, Paschim Medinipur, West Bengal, India, for his great encouragement in scientific endeavorance and whom she regards as an iconic person with incredible capabilities. PD is sincerely grateful to Emeritus Prof. O. N. Srivastava, Dept. of Physics, Institute of Science, Banaras Hindu University, Varanasi, India, for TEM images and EDX; and Prof. S. K. Sengupta, Dept. of Chemistry, Institute of Science, Banaras Hindu University, Varanasi, India, for the characterization of the thermal properties of the nanomaterials. PD is earnestly thankful to Prof. Rajiv Prakash, Dean (R&D), Indian Institute of Technology (Banaras Hindu University), Varanasi, India, for the advanced characterization of the nanomaterials at the Central Instrument Facility (CIF) of Indian Institute of Technology (Banaras Hindu University), Varanasi, India; Prof. A. S. K. Sinha, Dean (Academics), Indian Institute of Technology (Banaras Hindu University), Varanasi, India, for further characterizations; Prof. P. C. Pandey, Dept. of Chemistry, Indian Institute of Technology (Banaras Hindu University), Varanasi, India, for his research support and Prof. Shyam Sunder, Dept. of Medicine, Institute of Medical Science, Banaras Hindu University, Varanasi, India, for the cytotoxicity test of nanomaterials. PD is also grateful to Emeritus Prof. S. Pal, Dept. of Biomedical Engineering, Jadavpur University, India, for scratch test; Dr. M. M. Dwivedi, National Centre of Experimental Mineralogy and Petrology, University of Allahabad, Prayagraj, India, for SEM observations; and Mr. Shakti Nath Das, S. N. Bose National Centre for Basic Sciences, Kolkata, India, for EDX. PD is earnestly grateful to the faculty members, Prof. N. D. Pandey, Prof. P. K. Dutta, Dr. Tamal Ghosh and Dr. Ashutosh Pandey, of the Dept. of Chemistry, Motilal Nehru National Institute of Technology Allahabad, Prayagraj, India, for extending their cooperation. PD pays her sincere gratitude to the Directors, supporting staffs and her colleagues of Indian Institute of Technology (Banaras Hindu University), Varanasi, India, and Motilal Nehru National Institute of Technology Allahabad, Prayagraj, India.

PD acknowledges the research grant by the Science and Engineering Research Board (SERB), a statutory body of the Department of Science and Technology (DST), Government of India (GoI), through the Project File No.: PDF/2017/002264 under the National Post Doctoral Fellowship (N-PDF) scheme.

The authors acknowledge CRC Press/Taylor & Francis Group for granting approval for publishing the research monograph.

The authors pay their heartfelt homage to all the holy souls of the other world who made things in favor. They do believe what is rightly stated by Paulo Coelho, 'And, when you want something, all the universe conspires in helping you to achieve it'.

About the Authors

Dr. Poushpi Dwivedi obtained her BSc (Hons.) in Chemistry (2001), MSc in Chemistry (2003) from Banaras Hindu University, Varanasi, India, and PhD in Chemistry (2015) from Motilal Nehru National Institute of Technology Allahabad, Prayagraj, India. She has been a DST-SERB national post-doctoral fellow at the Department of Chemistry, Indian Institute of Technology (Banaras Hindu University), Varanasi, India (2017–2019). She has also worked as project fellow (2004) and post-doctoral fellow (2017) in the Department of Chemical Engineering & Technology, Indian Institute of Technology (Banaras Hindu University), Varanasi, India; as guest faculty (2016) in the Department of Chemistry, University of Allahabad, Prayagraj, India. She is presently an assistant professor of Chemistry in the Department of Chemistry (UG & PG), Belda College, Vidyasagar University, Paschim Medinipur, India. Her many publications consist of journal papers, book chapters as well as national and international conference proceedings.

Prof. Shahid S. Narvi is a professor in the Chemistry Department, Motilal Nehru National Institute of Technology Allahabad, Prayagraj, India, working as a faculty member for the last 34 years. He graduated with BSc (Hons.) in Chemistry (1977) and MSc in Chemistry (1979) from Aligarh Muslim University, Aligarh, India. He received his PhD in Chemistry (1985) from University of Roorkee, (Now IIT Roorkee), India. He has great number of publications with a lot of research and teaching experience together with supervising MTech and PhD research work.

Prof. Ravi Prakash Tewari is a professor in Applied Mechanics Department, Motilal Nehru National Institute of Technology Allahabad, Prayagraj, India, working as a faculty for more than 15 years. He graduated with a BE in Mechanical Engineering (1990) from Madan Mohan Malviya Engineering College, Gorakhpur, India. He received his MTech (1993) and PhD (1999) in Biomedical Engineering from Institute of Technology, Banaras Hindu University, Varanasi, India. Besides having myraid publications, he has extensive experience in research and teaching as well as in supervising MTech and PhD research work.

Prof. Dhanesh Tiwary is a professor in the Department of Chemistry, Indian Institute of Technology (Banaras Hindu University), Varanasi, India, working there as a faculty member for more than 16 years. He graduated with a BSc (Hons.) in Chemistry (1984), postgraduated with a MSc in Chemistry (1986) and obtained his PhD in Chemistry (1992) from Banaras Hindu University, Varanasi, India. He has more than 21 years of teaching experience with 7 years of national and overseas postdoctoral experience. He has numerous publications to his credit and extensive experience in supervising MTech and PhD research work

Abbreviations

AAm	Acrylamide
AAS	Atomic absorption spectroscopy
ACE2	Angiotensin-converting enzyme 2
ADV	Adenovirus
Ag⁺	Silver ions
Ag⁰	Silver atoms
Ag/CS	Silver/chitosan
Ag/CS-g-PAAm	Silver-g-poly(acryl amide)
Ag/CS/PVC	Silver/chitosan/poly(vinyl chloride)
AgNO₃	Silver nitrate
AgNP	Silver nanoparticle
APS	Ammonium persulfate
BAI	Biomaterials associated infection
BNC	Bionanocomposite
CaCO₃	Calcium carbonate
CaO	Calcium oxide
CoV	Coronavirus
α-CoV	Alpha-coronavirus
β-CoV	Beta-coronavirus
γ-CoV	Gamma-coronavirus
δ-CoV	Delta-coronavirus
COVID-19	Coronavirus disease 2019
CS	Chitosan
DNA	Deoxyribonucleic acid
DSC	Differential scanning calorimetry
EDX	Energy dispersive X-ray analysis
EETMOS	2-(3,4-epoxycyclohexyl)ethyltrimethoxysilane
EHS	Environmental Health and Safety
EPA	US Environmental Protection Agency
FE-SEM	Field emission - scanning electron microscopy
FDA	Food and Drug Administration
FTIR	Fourier transformed-infrared
GA	Glutaraldehyde
HE	Holy basil extract
HIV-1	Human immunodeficiency virus
HNB	Human finger and toe nail pruning biomass
H₂O₂	Hydrogen peroxide
HR-TEM	High resolution - transmission electron microscopy
HSV	Herpes simplex virus
IHD	Ischemic heart disease
K₂PdCl₄	Potassium tetrachloropalladate

LCA	Life cycle assessment
MBA	N,N'- methylenebisacrylamide
MERS-CoV	Middle East respiratory syndrome coronavirus
$Mg(NO_3)_2$	Magnesium nitrate
MgO	Magnesium oxide
MgO/CaO nanoalloy	Magnesium oxide/calcium oxide nanoalloy
MTT	3-(4,5-dimethylthiazol-2-yl)-2,5-diphenyl tetrazolium bromide
NaOH	Sodium hydroxide
NBNM	Nanobiotechnologically synthesized nanomaterial
NC	Nanocomposite
NDM-l	New Delhi Metallo-beta-lactamase
NH_4-Y	Ammonium Y-zeolite
NiV	Nipah virus
OD	Optical density
PAAm	Poly(acrylamide)
PBS	Phosphate buffer saline
Pd^+	Palladium cation
Pd^0	Palladium atom
PdNP	Palladium nanoparticle
PVC	Poly(vinyl chloride)
PVP	Polyvinylpyrrolidone
RBD	Receptor-binding domain
RNA	Ribonucleic acid
ROS	Reactive oxygen species
RSV	Respiratory syncytial virus
SAED	Selected area electron diffraction
SARS-CoV-2	Severe acute respiratory syndrome coronavirus-2
SEM	Scanning electron microscopy
SNNI	Safer Nanomaterials and Nanomanufacturing Initiative
SPR	Surface plasmon resonance
SR	Swelling ratio
TEM	Transmission electron microscopy
TEMED	N,N,N',N'-tetramethylethylenediamine
TG/DTA	Thermogravimetric/Differential thermal analysis
THF	Tetrahydrofuran
UV	Ultra violet
WCED	World Commission on Environment and Development
XRD	X-ray diffraction
ZOI	Zone of inhibition

1 Nanobiotechnology toward the Next Generation Antimicrobial Materials

1.1 NANOTECHNOLOGY: INNOVATION OF THE NEXT BIG TINY THING

Richard P. Feynman, Nobel Prize winner and a world-renowned US physicist, presented a speech entitled, *There's Plenty of Room at the Bottom* during the annual meeting of American Physical Society at the California Institute of Technology in 1959 [1, 2]. Feynman through his speech was first to articulate publicly the manipulation of matter on a tiny scale. Now, years later, significant advances in nanotechnology are transforming his historic speech much more into a reality, with myriads of applications in various arrays of life. Albeit, Feynman is credited for creating nanotechnology, Norio Taniguchi introduced the term 'nanotechnology' in the year 1974 for representing extra-high precision nanoparticles with ultrafine dimensions [3], which have been around us since ancient times.

The first evidence of nanotechnology dates back to ~2000 BC when silver and gold nanoparticles were used to create beautiful stained glass windows owing to their unique optical properties [4]. The Romans mixed solutions of gold and silver nanoparticles to produce different types of colored glasses, like ruby red, lemon yellow, etc. The Romans also believed that an *Elixir of Life* could be created using soluble gold and silver solutions, which could cure many ailments and may increase mental as well as physical abilities [5]. The first example to receive some recognition is the Roman Lycurgus Cup, a fourth-century AD bronze cup lined with colored glass. This glass transmits red light and scatters a dull green light. The British Museum, which currently displays the cup, has concluded that the glass contains ~70 nm-sized particles that are an alloy of (70%) silver and (30%) gold [6]. Silver nanoparticles having this size can scatter green light and transmit orange, while the addition of gold nanoparticles will shift the absorption band toward longer wavelength [7]. The Romans and Greeks, even in those days (2000 years ago) used sulfide nanocrystals to dye hair. Damascus steel or wootz, developed in India around 300 BC and exported to the Middle East during the medieval times, for the production of patterned swords used by the Muslims, are demonstrated to possess carbon nanotubes, formed by protecting nanowires of

DOI: 10.1201/9781003217343-1

1

cementite (Fe_3C), which were a hard and brittle compound from iron and carbon of the steel [8]. The blacksmiths of olden days were inadvertently using nanotechnology, and they carefully maintained this secret to their manufacture; though the secrecy and lack of transmission eventually led to ceasing out of this technique in the eighteenth century.

Today, researchers and scientists are using nanotechnology to knock out procrastinating challenges in almost every field, such as the sector of agriculture [9], food [10], energy [11], medicine [12], catalysis [13, 14], chemical and biological sensor analyses [15, 16], construction [17, 18], cosmetics [19], textile industry [20], environmental science [21], space research [22, 23], security [24], robotics [25], electronics [26], computers [27] and information technology [28], to mention a few.

1.1.1 PORTRAYING NANOTECHNOLOGY

The term nanotechnology is based on the prefix 'nano' originating from the Greek word *nanos*, which means dwarf or extremely small. In rather technical terms, 'nano' in SI units denotes to 10^{-9} m, equal to one billionth (0.000000001) of a meter (m) or one nanometer (nm). For comparison, it is the width of few atoms, and on the scale of individual molecules; a carbon nanotube is of 1 nm in diameter, a DNA molecule is approximately 2-nm wide, and virus is roughly 100 nm in size. Conceptually, nanotechnology is thus the creation of functional materials, devices and systems through the understanding, union and collaboration of different scientific disciplines like chemistry, physics, biology, computer and material sciences integrated with engineering to harness and control matter, typically, but not exclusively, between 1 and 100 nm, in any one or more dimensions, where the onset of size-dependent phenomena usually enables novel applications with transition in behavior [29].

1.1.2 CHARACTERISTICS OF NANOMATERIALS

At the nanoscale, matter demonstrates totally new and unique properties. It becomes stronger and better conductor of heat and electricity, with even superior biological characteristics. These different properties from bulk (micrometric or larger) materials arise as a result of their size. The classical laws of chemistry and physics cannot be readily applied at this very small scale. First, the electronic properties of very small particles are very different from their larger ones. Second, due to the high surface area to volume ratio, and since the surface atoms are generally most reactive, having the high energy due to unsatisfied valency, the properties of a material transform entirely in unexpected ways. Example, when silver turns into nanosized particles, it takes on bactericidal properties, while gold in macro scale is solid, inert and yellow at room temperature but starts changing color depending on the size. Gold at nanoscale becomes liquid and red in color, together with incredible catalytic properties unseen at macro scale [30]. Aluminum cane is safe and stable, but aluminum at nanosize is extremely explosive and combustible. Copper, an opaque substance, becomes transparent, while platinum being an inert material becomes a catalyst and silicon an insulator become conductor at nano level. Nature, too, provides plenty of nanotechnological properties: the sleekness of dolphin skin, the iridescence of butterfly wings, the 'nanofur' of gecko foot pad imparting strong adhesiveness that allows geckos to walk up on vertical surfaces, etc.

Therefore, the characteristic features of extraordinary physical, chemical, electrical, optical, magnetic and antimicrobial effects can be controlled through manipulating matter on an atomic or molecular scale, which may then be exploited higher at the micro or macro level for creation of structures and systems having deliberately engineered with unique properties, novel functionalities and enhanced performances [31]. Nanotechnology is a science inherently of multidisciplinary nature: drawing together diverse branches of basic sciences, applied streams, engineering and technologies to unify into a single domain at one point. The important point is that the integration is for the purpose of laying a scenario which over the coming years will provide unexpected advances across most existing research with a world of unforeseen development.

1.1.3 BREAKTHROUGHS AND OPPORTUNITIES

Nanotechnology evidently is to probate a profound effect on the global economy of the society encompassing every science and technology within the early twenty-first century. Nanotechnologists promise progressiveness with breakthrough achievements in areas of materials manufacturing, to medicine and healthcare. It is widely and worthfully felt that nanotechnology will be the next big tiny thing in the industrial and technological revolution.

There are major landmarks in the birth and life of nanotechnology. The scanning tunneling microscope (STM) was invented by Gerd Binnig and Heinrich Rohrer in 1981 for imaging at the nano range, followed by the invention of the atomic force microscope (AFM) by Binnig, Quate and Gerbein in 1986 for observing image structures at the atomic scale [32]. In 1985, Harry Kroto, Robert Curl and Richard Smalley brought a memorable turn by bringing forth a new form of carbon (carbon allotrope), the fullerenes. The most common fullerene, the Buckminster fullerene (C_{60}), called 'buckyball,' measures ~7 Å (0.7 nm) in diameter, resembles a soccer ball, with a single molecule of 60 carbon atoms arranged having 20 hexagons and 12 pentagons, symmetrically [33]. The Nobel Prize in Chemistry was also awarded for this in 1996. After fullerenes, we had the carbon nanotubes, discovered by S. Iijima in 1991. The first book on nanotechnology, *Engines of Creation,* by K. Eric Drexler was introduced to the world in 1986. In his book, future prospects of nanotechnology, opportunities and creations were discussed, particularly the idea of 'molecular nanotechnology' and 'bottom-up approach' which is the self-assembly of atoms and molecules to larger components, in an ordered fashion into a functional structure [34, 35].

1.1.4 APPLICATIONS AND CHALLENGES

The desire to acquire far-reaching goals in the foreground of already treading avenues with obstacles still left to overcome lay in: medical applications [36, 37], electronics and information technologies [38], energy production and storage [11], materials science [39], space exploration [40], manufacturing [41], instrumentation [42], food, water and environmental [43–46] and security [47].

Envisaging strategies via nanotechnologies hyped and in high demands can be briefed under the following heads:

- Sensors for biomedical and environmental and chemical monitoring purposes.
- Advanced materials for medical, defense, aeronautical and automotive industries.
- Instant mapping of genetic code.
- Lab-on-chip diagnostic techniques.
- Longer-lasting medical implants.
- Double the ability of survival and extend life span.

1.1.5 BIORESOURCE-ENABLED NANOMATERIAL SYNTHESIS: FUTURE

Nowadays, bioresources-enabled nanomaterial syntheses, such as plant materials (phytomass) and other biomass-involving microorganisms, are becoming extremely popular, which holds tremendous future applicability in the sector of nanomedicine, because of the green factor rendering superior biocompatibility. Most importantly, nanotechnology being a specialty for physicists, chemists and other scientists has now become a special tool for physicians, surgeons and other medical practitioners. Rather, 'nanomedicine,' is now the well-defined term for applying the principles of nanotechnology and use of engineered nanostructures and nanodevices for repair, construction and monitoring in the domain of disease diagnosis, treatment and health care [48].

Nanomedicine is therefore a wedge-driven part of nanotechnology, which comes into being based on complete knowledge of human body concurrently with concepts to produce nanomaterials designed to interact with biological systems and destined to bring radical transformation in the traditional doctor-patient relationship. In other words, nanomedicine, which is human body and disease centered, is recently trying to do better, bring perhaps more profound transformations of health care on a molecular level as well as on the nanoscale. Examples of bewildering nanomedicine in action are nanoscale materials for scaffolding, diagnostic tools, biosensors, etc.; biocidal nanoparticles giving rise to antimicrobial surfaces for minimizing infection in implanted biomaterials, surgical devices, doctors' offices and hospital buildings [49]. In the platforms of physicochemical functionalization, biodistribution and pharmacokinetics nanotechnological characteristics prove to be highly specific and provide benefit not achieved previously. In the book, *Nanomedicine*, the first book on nanomedicine in 1999, Robert

Freitas assembled the most balanced view point of nanomedicine, with present and future perspectives, bringing larger insight into its developmental aspects and beliefs to consider for even today [50].

1.1.5.1 Applications of Nanomedicine

The nanomedical sector based on its applications can be put into dignified sections, which include *in vivo* imaging, diagnostics, therapeutics, drug formulations, drug delivery, preventive nanomedicine, regenerative medicine, active implants and biomaterials. Development oriented active research in this area is currently especially concentrated in the domains involving: tissue regeneration – regenerative medicine or tissue engineering toward body's self-repair [51], cancer therapy [52], toward the cure of HIV/AIDS [53], fullerenes [54, 55], diagnostics – toward personalized medicine, validating intelligent, automated diagnosis and realize truly the dream of tremendously prospected 'lab on a chip' perspective [56].

1.1.5.2 Present and Future Scenario of Nanomedicine

The fruitful services of nanotechnology in the nanomedicine platform can be briefly considered brought about by nanoparticles for implant surfaces, drug carriers, tissue engineering and nanoengineered devices for diagnostic and biosensing applications [57]. Henceforth, physicians, surgeons, medical practitioners, biomedical engineers, scientists and nanotechnologists identified several key topics for nanomedicine research in the coming years: applications of nanotechnology in therapeutics, diagnostics, instrumentation, tissue engineering, developing nanostructures having antimicrobial surfaces, with focusing the innovation on biological and biomimetic nanostructures, together with an understanding of biological-nanomaterial interface [58].

It is further speculated that aspects and modalities of nanomedicine with concurrent applications would possibly play an important role in the advancement of bioengineering realizing the promise associated with the term 'nanomedicine.' One of the main future aims of nanomedicine revolves around 'polymer therapeutics' (rational design of nanomaterials in nanomedicine). Obviously, there is an oversight that the utility of nanomedicine is likely to extend enormously through engineered polymeric nanomaterials into diagnostics, therapeutics, molecular research techniques and tools, explicitly with adequately evaluated safety factors to raise the 'quality of life' globally. Nanomedicine is evidently going to be the 'holy grail' of health care in the near future.

1.2 OUTLINING NANOMATERIALS

Starting with nanomaterials and nanostructures needs explicitly explaining nanoscience and nanotechnology separately. The subject that deals with the understanding part of the new field and understanding the fundamental physical and chemical properties of the nanostructures is nanoscience. While nanotechnology is the subject that deals with the applied part of the new field. Application of the understanding of the basic principles in the control and manipulation of nanostructures and their synthesis for the preparation of macroscopic structures having the desired properties is nanotechnology. Science and technology both converge together for the synthesis, characterization and classification of nanomaterials. Nanoscience and nanotechnology, in a broad sense, are involved in the engineering of nanomaterials, having one or more dimensions on the nanoscale, ranging from a few to even several hundred nanometers.

1.2.1 NANOMATERIALS

The 2011 Commission Recommendation defines 'nanomaterial' as, *a natural, incidental or manufactured material containing particles, in an unbound state or as an aggregate or as an agglomerate and where, for 50% or more of the particles in the number size distribution, one or more external dimensions is in the size range1 nm–100 nm. In specific cases and where warranted by concerns for the environment, health, safety or competitiveness the number size distribution threshold of 50% may be replaced by a threshold between 1 and 50%* [59].

1.2.2 CLASSIFICATION OF NANOMATERIALS

Nanomaterials are generally classified based on dimension of the materials being developed as well as on the phase composition [60].

Classification of nanomaterials dimension wise: Based on the number of external dimensions of the material, outside the nanorange (i.e. nanoscale ranging from 1 to ≤ 100 nm)

- Zero dimensional (0D) = all the dimensions of which are within the nanorange ~ (1–100) nm (and no dimensions >100 nm)
  e.g. nanospheres
- One dimensional (1D) = one dimension of which is outside the nanorange
  e.g. nanorods, nanofibers, nanowires, nanotubes, etc.
- Two dimensional (2D) = two dimensions of which are outside the nanorange
  e.g. graphene, nanoplates (plate-like shapes), nanoplatelets, nanofilms, nanolayers, nanocoatings, etc.
- Three dimensional (3D) = class not confined to the nanorange in any of the dimension
  e.g. multi-nanolayers, dispersions of nanoparticles, bundles of nanotubes, nanowires, etc.

Classification of nanomaterials based on phase composition:

- Single-phase solids = amorphous, crystalline particles, etc.
- Multi-phase solids = coated particles, matrix composites, etc.
- Multi-phase systems = ferrofluids, aerogels, colloids, etc.

TABLE 1.1
Descriptions of 'Nano' Suffix Terms [60]

Type	Description
Nanoscale	Size range of approximately 1–100 nm
Nanoparticle/Nano-object	Material: one, two or three external dimensions in nanoscale
Nanosphere	Nano-object having all the three external dimensions in nanoscale
Nanofiber	Nano-object (flexible/rigid) having two external dimensions in nanoscale while the third dimension being significantly larger
Nanorod	Solid nanofiber
Nanotube	Hollow nanofiber
Nanowire	Electrically conducting or semiconducting nanofiber
Nanoplate	Nano-object having at least one external dimension in nanoscale while other two external dimensions being significantly larger

Source: Adapted from ISO/TS 27687.

Though the fundamental properties of nanostructures differ from their bulk macroscale counterparts, they allow control over the physical and chemical properties at the macroscopic level as well. Therefore, nanotechnology provides the opportunity to prepare nanomaterials, which represent potentially created better materials and products. Table 1.1 and Table 1.2, provide recommended terms and scale limits regarding nanomaterials.

1.2.3 NANOBIOMATERIAL SYNTHESES

There are two main routes for creating nanomaterials and nanostructures: the 'top-down' approach, which involves miniaturization of macro- and microsystems; and

TABLE 1.2
List of Recommended Upper Limits by Different Organizations [60, 61]

Upper Limit (nm)	Source
100	ASTM
100	ISO
100	Royal Society – SCENIHR
100	Swiss Re
100	ETC group
200	Defra
200	Soil Association
300	Friends of Earth
300	Chatham House
500	Swiss federal Office of Public Health
1000	House of Lords Science Committee

the other one is the 'bottom-up' approach that develops nanostructures starting from atomic and molecular level [61]. This latter bottom-up approach can be associated with assembly or to synthesis, whereby nanostructures are formed atom by atom, molecule by molecule, using physical or chemical processes or spontaneous self-assembly, and is currently considered as the 'embryo of nanotechnology.' By contrast, the former, top-down technique is like milling or sculpting from any base material involving the steps of patterning and etching. Nanoscale fabrication through top-down approaches include optical together with scanning probe lithography, atomic force microscopic lithography, material removal and deposition (chemical, mechanical or ultrasonic), attrition, repeated quenching, printing and imprinting, etc. While bottom-up approaches include layer-by-layer self-assembly, molecular self-assembly, direct assembly, coating and growth, colloidal aggregation, etc. Examples of nanomaterial-containing products already available in the markets include food, home accessories, scratch-free paints, coatings, clothing, cosmetics, computers, electronic components, sports equipment, health and fitness devices, etc.

1.2.4 NANOPARTICLES

An aggregate of arbitrary $10-10^5$ atoms bonded together and having a radius of around (1–100) nm is defined as a nanoparticle. Due to the extremely small size of nanoparticles, the ratio of surface area to volume is much high and because of the electrons being compressed into a small area which give rise to 'quantum confine-ment effect' are emerging several new properties. Over the past decades, significant advances have been made in this field exploring these characteristics; particularly for metal nanoparticles, whose nanometer-scale features are mainly built up from their elemental constituents. Silver and gold nanoparticles have unique size- and shape-dependent properties (optical [62, 63], chemical [64], photothermal [65] and catalytic [66, 67]), making them useful for a variety of applications. Their optical properties are result of the collective excitation of free electrons in response to light energy; this phenomenon is commonly known as surface plasmon resonance (SPR) [62]. The plasmon resonance of silver and gold nanoparticles is readily tun-able with the nanoparticle size, shape, surface chemistry and surrounding media (refractive index) – making them ideal for sensor applications [68, 69]. Silver also has well known other biological activity; therefore, their nanoparticles are immensely valued in the medical field holding innumerable applications. The syn-theses of metal nanoparticles are discussed in a broad range of articles and it is realized that these are achieved through a variety of processing techniques. A wide range of techniques are also used for their characterization [70–72].

1.2.5 NANOCOMPOSITES: POLYMER NANOCOMPOSITES
AS THE ADVANCED COMPOSITES

Composites are formed by the judicious combination of two or more materials, producing a synergistic effect and taking into advantage the salient features of each of the materials. They are composed of two or more physically distinct phases: one being typically the continuous phase (matrix) and the other one or more being the discontinuous phase (reinforcement). There is presence of an interphase, and the properties shown by the composite materials differ from the initial materials. Typical examples of natural composites are bone and wood; bone having hydroxy-apatite reinforced in collagen and wood with cellulose, hemicellulose and lignin.

Composites can be classified on the following several bases:

1. Function: electrical and structural.
2. Geometry of reinforcements: fiber composites and particulate composites.
3. Matrix: metal matrix composites (MMC), ceramic matrix composites (CMC) and polymer matrix composites (PMC) [73, 74]. PMC are the highly coveted composites having the reinforcement of polymers with fillers whether organic or inorganic.

With the advent and application of polymer composites, chiefly in the nanomedi-cine landscape, a new era with unique opportunities to create revolutionary mate-rials by exploring the fundamental chemistry had proceeded. Many researchers are investing time to modify the properties of conventional polymers by incor-porating nano-fillers for improved biocompatibility and functional behavior. A new class of composite materials now have been successfully developed with the combination of organic polymer matrix and inorganic nanoscale fillers, in which the dispersed particles have one or more dimensions in the nanoscale range (1–100 nm), called polymer nanocomposites. Polymer nanocomposites are the most advanced type among all the assorted materials with many end applications.

1.2.5.1 Properties of Polymer Nanocomposite

These exhibit marked extraordinary properties with improved characteristics in comparison to the pure polymers or ordinary composites. To account a few are: (1) enhanced strength and modulus, (2) better barrier qualities, (3) improved resis-tance to heat and solvent, (4) reduced flammability, (5) novel optical proper-ties, (6) variation in T_g values of the polymer, (7) overall changes in physico-chemical properties, (8) enhanced biocompatibility and (9) raised bactericidal activity [75–81].

1.2.5.2 Fabrication of Polymer Nanocomposites

The synthesis of polymer nanocomposites is an integral aspect of polymer nano-technology. The assembly and the interactions of molecular or polymeric spe-cies with the inorganic nano-substrates constitute the basis for the preparation of nanocomposites. The steps employed are: (1) polymer selection and modification for the matrix (continuous) phase, (2) choice of a biocompatible nanostructured reinforced (discontinuous) phase and (3) dispersion of the discontinuous phase in the continuous phase. Dispersing the discontinuous phase – nanoparticles in the polymer matrix can be made possible via two ways: the first kind is by mixing the nanoprecursor to the polymer matrix and then allowing *in situ* formation of the particles, the second way is the 'Breath in and breath out technique' – the particles are made to form separately and then reinforced in the polymer matrix.

The chemical and the physical interactions between the particulate and the polymeric phases effectively stabilize the nanoparticles and prevent agglomeration or clustering [82].

Nanosized particles usually used for the preparation of inorganic particle/polymer nanocomposites are: metals (Ag, Au, Cu, Zn, Ti, Pd, Pt, Al, Fe, etc.) and some of the metal oxides (Ag_2O, CuO, ZnO, Al_2O_3, $CaCO_3$, TiO_2, etc.) [83]. Different procedures for forming nanocomposites, popularly implied include: (1) solvent casting, (2) blow and injection molding, (3) electrospinning, (4) film blowing, (5) foam extrusion, (6) pultrusion, (7) autoclave technique, (8) layer-by-layer assembly, (9) wet lay-up, etc. [84, 85].

1.2.6 TECHNIQUES OF CHARACTERIZATION

Characterization is an important aspect of dealing with materials, which contains all those features required to reproduce the material. The essential parts of characterization are investigation of chemical composition and structural identification. This may also comprise analysis of impurities, defects, chemical, and physical homogeneity affecting the material properties [86, 87].

UV-visible spectroscopy is an effective tool for studying the optical properties of nanomaterials, which facilitates in defining the SPR, exhibited by different nanostructured particles at specific wavelength. The origin of SPR is due to the collective oscillations of electron density or slight electron drift with any interacting electromagnetic radiation. These types of resonances are termed as the surface plasmons. ELICO SL 159 version of UV-Vis spectrophotometer is displayed in Figure 1.1.

FIGURE 1.1   UV-Vis spectrophotometer instrument.

Transmission electron microscopy (TEM) is an advanced technique to visualize the morphological structure of the nanomaterial. Deep observation is possible from different orientations, locations and at different magnifications. TEM images provide a representative picture to study and establish the physical features of the nanomaterial. HR TEM FEI Type: TECHNAI G2 20 TWIN, instrumental set-up attached with an energy dispersive X-ray spectroscopic analysis (EDAX) system is presented in Figure 1.2. HR-TEM image of palladium nanoparticles is illustrated in Figure 1.3.

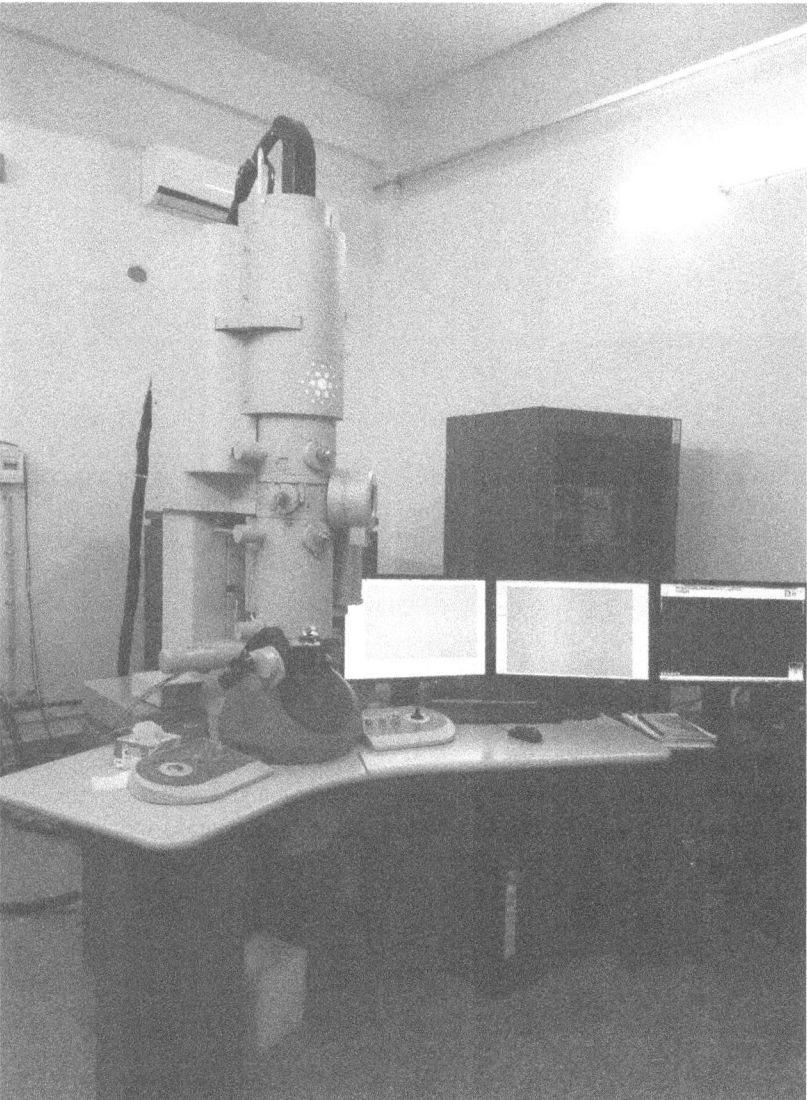

FIGURE 1.2   HR-TEM instrumental set-up attached with EDAX.

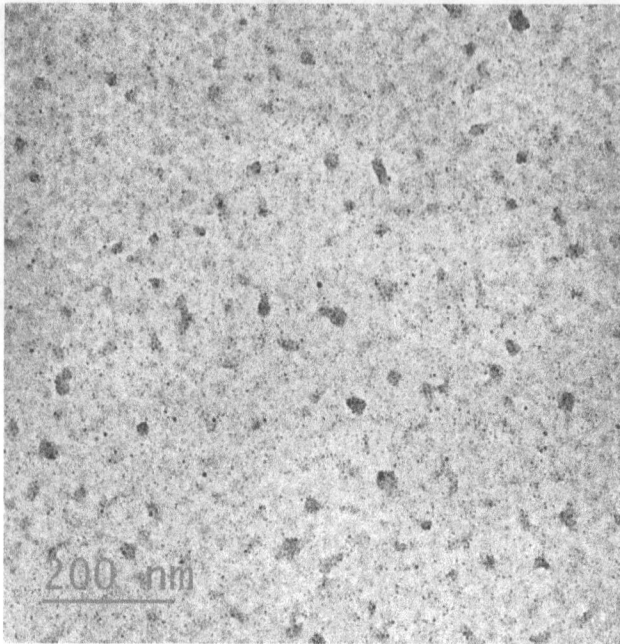

FIGURE 1.3   HR-TEM image of Pd nanoparticles.

Scanning electron microscopy (SEM) offers an overview of the morphological structure at the nanoscopic level and is a useful tool for the characterization
and elucidation of nanoscale materials. FBI Nova NanoSEM 450, FE-SEM instrumental set-up, attached with an EDAX system is presented in Figure 1.4. SEM
image of silver nanoparticles is illustrated in Figure 1.5. Few other microscopies,
such as scanning tunneling microscopy (STM) and scanning probe microscopy
(SPM) like atomic force microscopy (AFM) play a vital role in determining
the structural properties and phenomena of the nanomaterials. Scanning probe
microscope NT-MDT instrument is shown in Figure 1.6.

X-ray diffraction (XRD) helps in identifying the nature of bonding, order of
crystallinity and distinguishing between crystalline and amorphous substances.
XRD pattern of a crystalline material shows high sharp peaks in contrast to an
amorphous polymer which does not show any intense peak whereas polymeric
nanocomposite shows some extent of peak. This is largely due to the development
of crystallinity in the material possibly conferred by the crystalline nature of the
dispersant. Photograph of Rigaku MiniFlex600, XRD instrument, is shown in
Figure 1.7.

Energy dispersive X-ray spectroscopic analysis (EDAX, EDX or EDS) is
an analytical technique for the chemical characterization and identification of the
elemental composition of the nanomaterials. EDAX system identifies the specimen of interest through the imaging ability of the microscope and as attachments
of TEM or SEM instruments.

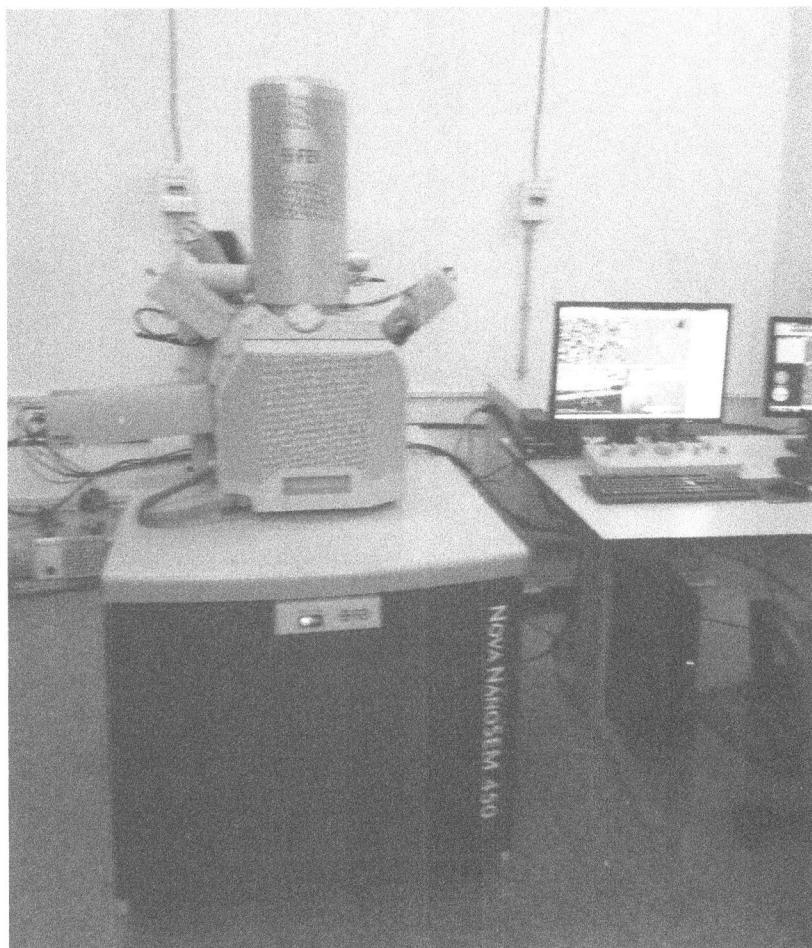

FIGURE 1.4 FE-SEM instrumental set-up attached with EDAX.

Fourier transform infrared (FTIR) spectroscopy is a very important technique for chemical composition analysis of materials. It provides information of the chemical and structural changes brought about and evaluates the structure of molecules through the presence of characteristic functional groups and bonds in a substance. FTIR spectrum, which shows the presence of nanoparticles as well as the polymer matrix of a nanocomposite, is therefore a fingerprint for its identification. Photograph of Thermo SCIENTIFIC NICOLET - iS5 FTIR spectrophotometer is shown in Figure 1.8.

Thermal analysis (TGA, DTA and DSC) basically measures, as a function of temperature, the physical and chemical aspects of materials. (1) Thermogravimetric analysis (TGA) records the weight change of a test sample as a function of temperature or even time; (2) the differential thermal analysis (DTA) measures the

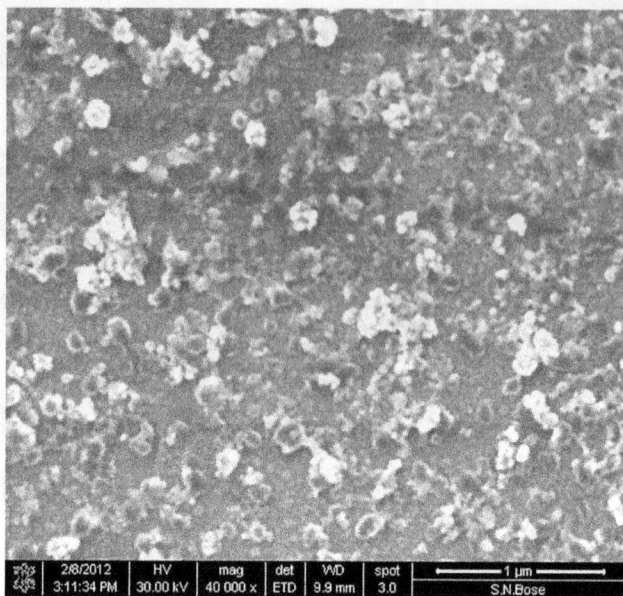

FIGURE 1.5 SEM image of Ag nanoparticles.

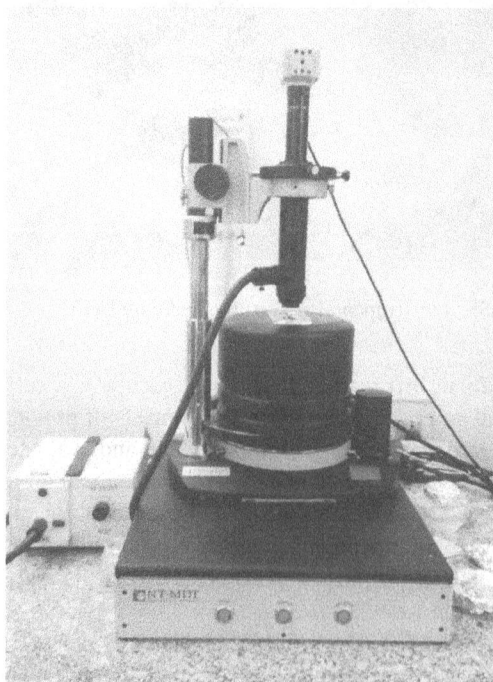

FIGURE 1.6 AFM instrument for characterization of materials.

FIGURE 1.7 XRD instrument for characterization of materials.

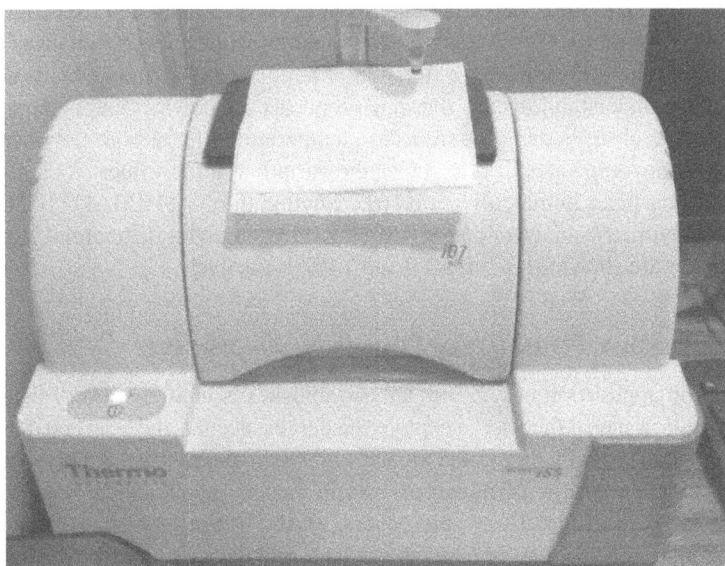

FIGURE 1.8 FTIR spectrophotometer for chemical composition elucidation of materials.

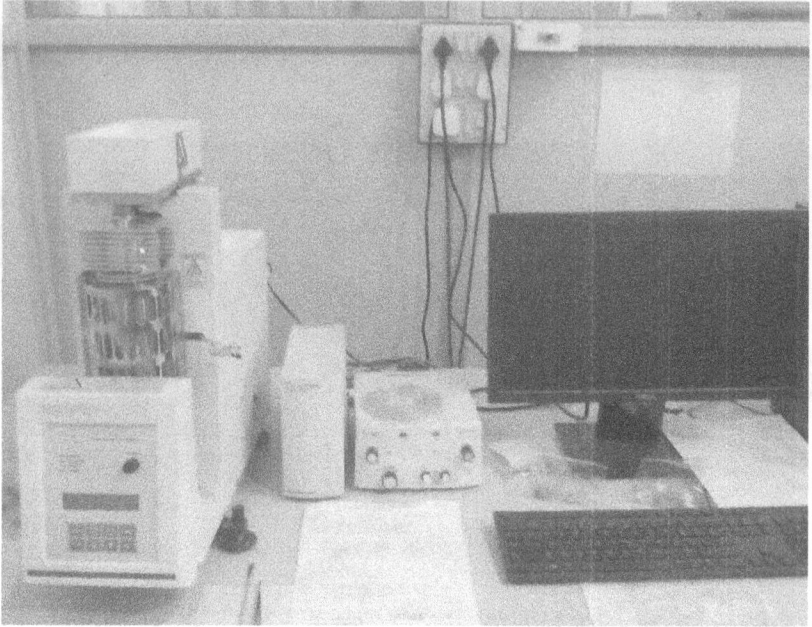

FIGURE 1.9 TGA instrument for determination of thermal transformation characteristics of materials.

temperature difference (T), between the test sample and any inert reference, as a function of temperature, detecting change in the heat content and (3) differential scanning calorimetry (DSC) quantitatively measures the enthalpy changes (ΔH), occurring in the test sample as a function of temperature or time. DSC is a helpful tool for understanding the thermal properties of polymer nanocomposites. DSC helps to observe the glass transition temperature (T_g) of both polymers and polymer nanocomposites. It also shows the variation in the values of (T_g), which represents particles in the polymer matrix. Photographs of SHIMADZU TGA-50 thermo-gravimetric analyzer and SHIMADZU DSC-60 Plus differential scanning calorimeter are shown in Figures 1.9 and 1.10, respectively.

1.2.7 Trending Potentials of Polymer Nanocomposites

Advent and application of polymer nanocomposites is in almost every field; few decades ago, it paved the way to explore the darkest avenue and addressed the far-reaching goals, specifically in the arena of biomedical sciences. Nanocomposites can redefine the sector of nanomedicine in various applications, e.g. in scaffolding, targeted drug delivery, biosensors, wound healing, prosthetic materials, catheter materials, coating of medical implants and devices, etc. Above all, nano-biomaterials through their superior performance can ameliorate the face of existing biomaterials in biomedical engineering.

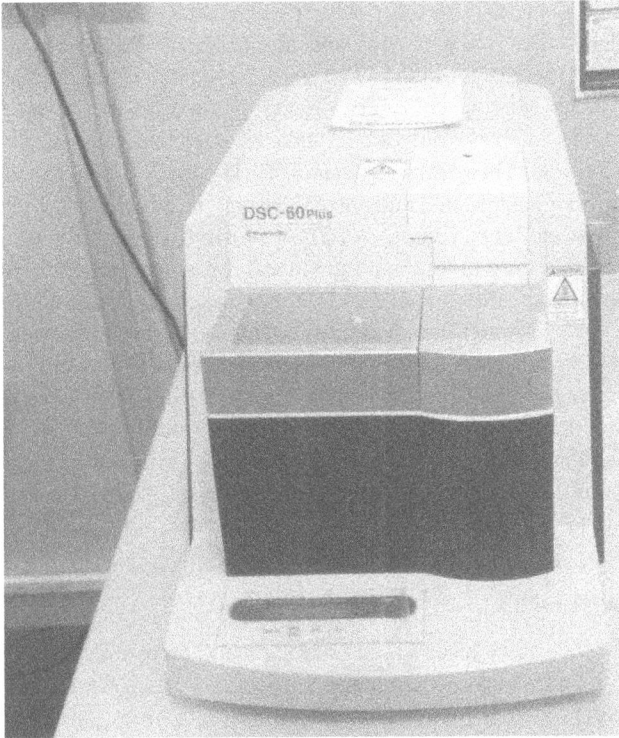

FIGURE 1.10 DSC instrument for determination of differential thermal properties of materials.

1.3 POLYMER NANOCOMPOSITES AS THE NEXT GENERATION BIOMATERIALS

Where the focus of nanotechnology in medical field is termed as 'nanomedicine'; at the same place, applying engineering principles, designing skills and techniques to medical and biological sciences for improving disease diagnosis, monitoring, therapy and health care can be called biomedical engineering. It consists of following domains – biomaterials, bioinstrumentation, imaging, biomechanics, bioinformatics, system engineering, system modeling, tissue engineering and clinical engineering. A confluence of multidisciplinary field is once again created for the domain of biomaterials, where by bioengineers, material scientists, immunologists, biologists, surgeons, chemists and nanotechnologists merge together to evolve strategies to develop novel biomaterials and deliver new nanotechnology-based medical tools and devices for diagnostics and therapeutics. A biomaterial can be defined as any substance which has been deliberately engineered for taking form alone or only as a part of a system, to direct a particular therapeutic or diagnostic procedure, by the control of some interactions with the components of the biological systems [88].

Potential applications of the polymer nanocomposites arising from different chemistries, their compositions and constructions, together with the interaction between the chosen matrix and the filler holds critical importance, as a biomaterial. The existing polymers preferably used in the biomedical arena [89] include the aliphatic polyesters, such as poly(caprolactone) (PCL), poly-lactide (PLA), poly(butylene succinate) (PBS), poly(p-dioxanone) (PPDO), poly(hydroxyalkanoate)s, etc., and natural biopolymers, such as starch, cellulose, chitin, chitosan, proteins, lignin, etc. The nanosized fillers utilized to fabricate the nanocomposites are inorganic, organic and metal nanoparticles, such as clay, magnetite, hydroxyapatite, nanotubes, chitin whiskers, lignin, cellulose, Cu, Zn, Au, Ag, etc. [90]. Though nanocomposite materials are taking roots in a variety of diverse health-care applications as biomaterials, there are knocking key research challenges in the development of self-sterilizing polymer nanocomposite materials for potent application in this area. Despite sterilization and aseptic procedures, microbial colonization causing nosocomial infection remains a serious complication in surgery and a setback in the utility of biomaterials like sensors, prostheses and implants.

1.3.1 Major Issues

Even after ~60 years of research and development, there still remain staggering issues regarding the use of biomaterials, such as IOLs (25% reoperation rate), hip prostheses (10–15 years lifetime), dental implants (infection), vascular grafts (no endothelialization), heart valves, catheters (90,000 deaths/year), cell culture surfaces (cell phenotype drifts), biocompatibility (no useful definition), etc. Major issues that continue to impede the performance of today's biomaterials are: (1) infection, (2) biostability, (3) healing and (4) blood clotting [91].

1.3.2 Infection Control

Biomaterials associated infections (BAI) is the second main factor of implant failure, coming right after the cause of instability [92]. Nosocomial infections on the surface of biomaterials are one of the chief barriers to long-term successful biomaterial integration. Infections are mainly caused by biofilm-forming bacteria, which are highly resistant to antibodies of the host defense mechanisms, antibiotics and other antimicrobial agents, as slimy protein matrix secreted by the organisms hinders diffusion or penetration into it. Microbes from the patient's own skin or mucosa can possibly infect during the incorporation of a medical device or implant. Given the very large population of patients with improved quality of life through biomedical materials, implants and devices, even a low risk of infection must be considered highly significant due to serious consequences. At the dawn of the twenty-first century, more than 2 million cases of hospital acquired infections (HAI) are reported annually in the United States alone and over half of these are associated with medical implants including catheters, prosthetics and subcutaneous sensors [93]. And now, with the recent arrival of multi-drug-resistant 'super-bugs' posing a serious threat, the figure of HAI is further expected to rise. BAI are thus a major bewildering clinical problem, which requires serious research attention and complete investigation.

The issues of BAI are briefed here: 2 million US patients/year acquire infection in the hospital; 90,000 patients die each year from infection in hospitals; hospital infection is in the top ten infectious diseases; $30,000 or more added to the hospital bill per infection; 250,000,000 catheters used per year in the United States, 5–10% have colonization; 1000× antibiotic dose required to kill biofilm bacteria. The global demand for significant renovation in this area has led to several preliminary strategic approaches and elementary diagnostic studies in this direction [94].

1.3.3 PATHOGENESIS AND DIAGNOSIS OF INFECTIONS

The interstitial milieu surrounding biomedical materials represent an area of local immune depression and a *locus minoris resistentiae*, known as immuno-incompetent fibro-inflammatory zone [95], which is prone to microbial colonization and infections. Moreover, the micro-movements of biomedical implants and prosthetic materials inserted into the body can damage the surrounding tissues and create conditions which deplete the immune defense [96–98]. Upon insertion of biomedical materials, there arises a competition between integration of the biomaterial to the surrounding tissue and racing microbes to grab the biomaterial surface. The presence of immune depression at the region of biomaterial implant surface-tissue interface, establishment of an infection is facilitated [99, 100].

It has been diagnosed that most of the BAI are chiefly caused by Gram-positive bacteria, mainly through staphylococcal species [101]. *Staphylococcus aureus* and *Staphylococcus epidermidis* together contribute to about 66% of orthopedic implant infections, while *Staphylococcus hominis* and *Staphylococcus haemolyticus* account for additional 13% whereas 22% is due to other pathogenic microbial species and 16% of them are polymicrobial in origin [102].

Thus, the understanding of the mechanism of pathogenesis by staphylococci [103] can help to deal with the problem of BAI in clinical and perioperative situations. Simultaneously, there lies the urgent need for developing self-sterilizing nanocomposite materials, which could serve positively like a 'gold standard' on the prevention level, to successfully compete the race between microbial adhesion and tissue integration.

1.3.4 STRATEGIES TO MINIMIZE INFECTIONS

Regarding BAI, although there were many established protective strategies, which included systemic perioperative antimicrobial prophylaxis through the administration of antiseptics and antibiotics, they are no longer much reliable [104, 105]. Several other infection-controlling strategies have been proposed [106], which are: (1) controlled release of antibiotics, (2) controlled release of some antibodies to resist microbial adhesion, (3) modification of surface topography, (4) nitric oxide release coatings and (5) coatings with nanostructured ZnO, TiO$_2$ and Ag rendering self-sterilization.

1.3.4.1   Coatings with Nanostructured ZnO, TiO$_2$ and Ag

This approach is used for coating of biomaterials especially with ZnO, TiO$_2$ or Ag nanoparticles, which are capable to release metal ions [107]. Exploitation of particular nanostructured materials for coating of medical implants and development of self-sterilizing biomaterials offers the modification of surface properties for achieving improved performance and biocompatibility, while at the same time decreasing bacterial adhesion. A study in which adhesion of S. epidermidis on nanostructured ZnO, TiO$_2$ and Ag was done to compare with microstructured surfaces proposed that nanophase materials were able to minimize S. epidermidis adhesion and boost osteoblast functions required to facilitate the efficacy of orthopedic biomedical implants [108]. In the case of titanium dioxide nanoparticles, the bactericidal effect was reached upon photoactivation with ultraviolet light [109]. On the other hand, silver nanoparticles, besides their wound healing ability, also exhibit high antibacterial activity even at very low content and assure this property for long term [110, 111], which makes them potential candidates having good prospects for use in the development of infection resistant next generation biomaterials.

1.3.5   ANTIMICROBIAL SILVER/POLYMER NANOCOMPOSITES

Conventionally, composites consisting of inert substrates and coated with metallic silver were found to be highly potential as a source of silver ions for controlling infection [112]. The effectiveness of silver-coated poly (ethylene terephthalate) for resisting bacterial colonization and subsequent biofilm formation was reported by Klueh et al. Low bacterial growth shows the inhibitory effect of silver ions (Ag$^+$), released from the material [113, 114]. Electrochemical deposition of silver from solution is another method of coating silver onto a substrate. However, this method suffers the drawbacks of low silver pickup and needs special technique for surface preparation [115]. Another convenient approach for obtaining antibacterial polymer composites is by incorporating silver into molten polymers [116]. In vitro antimicrobial activity against multi-resistant bacteria and in vitro cytotoxicity of nanosilver-filled PMMA bone cement that completely restricted bacterial proliferation were studied by Alt et al. [117].

Actually, silver-based antimicrobial materials seize great attention due to the non-toxicity of the active Ag$^+$ to human cells [118, 119] as well as their novelty of acting as a long lasting biocide together with being non-volatile and thermo stable. Ag$^+$ have strong antimicrobial properties and only a few bacteria may be resistant to this metal, which has been well established [120, 121]. Hence, silver is a significant resource for therapy and used in biomedical devices and implant materials [122, 123]. The antimicrobial action of silver depends on the Ag$^+$ released from silver containing materials, which binds strongly to electron donor groups like nitrogen, oxygen or sulfur, present in biological molecules. So, silver-based polymer nanocomposites should release the Ag^+ to the surrounding pathogenic environment in order to become effective. Metallic silver will be oxidized to the active Ag^+ possibly through interacting with the water molecules. Therefore, basic logic behind the selection of polymers, for the development of silver/polymer nanocomposites, should be the hydrophilic characteristic, which is required for steady Ag^+ release from the nanocomposite materials, to ensure proper and prolonged antimicrobial effect. The key factor for designing this class of materials is prolonged and steady discharge of the Ag^+ biocide at around concentration level of 0.1 ppb, which is capable of imparting an antimicrobial efficacy [124].

When silver nanoparticles are integrated into a polymer matrix to form nanocomposite or even if they are in combination with other metal nanoparticles forming bimetallic nanoparticles, combined desired properties are achieved. With each novel combination and integration of silver nanoparticles into nanocomposites, silver is explored and utilized to generate materials with new characteristics [125, 126].

1.4 SILVER

Silver is considered to be the 63rd metal most abundant in Earth's crust, and exists in the form of two stable isotopes, ^{107}Ag and ^{109}Ag, almost occurring in equal proportions. There are also 20 other radioisotopes of silver, none of them occurs naturally [127].

1.4.1 ELEMENTAL SILVER: CHARACTERISTICS AND SOURCES

Silver is a lustrous white transition metal element with the chemical symbol 'Ag' (Latin word: *argentum*, derived from the Indo-European origin *arg*–for 'white' or 'shining') and having the chemical properties; atomic number: 47, atomic weight: 107.868, melting point: 962°C, specific gravity: 10.5, hardness: 2.5. Silver and silver extraction have been known to the world since centuries. Man could separate silver from lead even during the early 3000 BC as indicated by the slag dumps in Asia Minor and on islands in the Aegean Sea. The metal is present naturally in the pure free form, i.e. native silver, as an alloy with other metals like gold in principal mineral sources and also in the form of ores, such as argentite (Ag_2S) and chlorargyrite ($AgCl_2$) [128].

Silver is classified with the precious metals because of its positive values and rarity. It is also a noble metal due to its resistance toward corrosion and oxidation, though not as much as gold, while slightly harder, little more abundant and much less expensive than gold. Silver is stable in normal air and pure water, but tarnishes when exposed to ozone, hydrogen sulfide or sulfur due to the formation of a black layer of silver sulfide. Among all the metals, silver has the highest electrical and thermal conductivity (the non-metal diamond and super fluid helium (II) have higher), lowest contact resistance and also highest optical reflectivity (though it is a poor reflector of UV light). It also has good malleable property and is very ductile. Silver is ideal for electrical applications but because of its higher cost and tarnishing tendency, it is not widely used and instead copper is used for electrical purposes. Above its boiling point, silver gives off green vapor. Silver has solubility in nitric acid and in hot sulfuric acid. Silver can be turned into powder, paste, flakes, printable sheets, wires, colloidal suspension and may be alloyed with other metals or converted into salts [129, 130].

1.4.2   SILVER: CHEMISTRY AND VARIED APPLICATIONS

The chemistry and applications of silver are being much studied after 1980, and current research has recognized that Ag^+ is very reactive in nature, being able to form different organic and inorganic complexes [131]. Silver is commonly in +1 oxidation state, e.g. silver nitrate ($AgNO_3$), also shows up in compounds with +2 and less commonly +3 oxidation states, e.g. silver (II) fluoride (AgF_2) and potassium tetrafluoroargentate ($K[AgF_4]$), respectively. Ag (II) complexes are not very stable in comparison to that of Ag (I) and Ag (III), because of inert pair effect. Ag^+ can easily bind to proteins in the biological systems and interact with trace metals present in our metabolic pathways.

Silver salts have outstanding properties, silver iodide (AgI) have the potentiality to produce rain by seeding clouds, silver chloride (AgCl) is used as pigment for glasses, while silver sulfide (Ag_2S) is used for metal work and in circuits. $AgNO_3$, the most important silver compound, which is transparent white crystalline solid, has high water solubility and photosensitivity. $AgNO_3$, is a starting reagent for many crucial syntheses and processes, it can also readily react to produce other silver compounds. This silver salt is used as a chemical for silver-plating, as indelible inks for stained glass, in photography and as an antiseptic in medicine. The application of $AgNO_3$ in medicine as an antiseptic agent dates back to 702–705 AD, first reported by Gabor [132].

New innovations and applications are continuously erupting exploring the remarkable properties of silver, for research and industrial use, apart from its illustrious reputation in jewelry, coins and silverware which will maintain the status of silver regarded as wealth and prestige. Silver has significant applications far and wide, it includes in the sector of energy, engines, electronics, electricals, printed circuitry, production of chemicals, photography, electroplating, brazing, soldering, mirrors, coatings, water purification, food, hygiene, and holds majestically a major contribution in pharmaceuticals and medicine.

1.4.3   THERAPEUTIC HISTORY DOWN THE AGES

Silver has for many years long been known to possess antifungal, antiviral and antibacterial properties, having broad spectrum activities [133–138] and is being safely used in the field of medicine [139–145] since 1800s [146, 147]. The Phoenicians even stored water, wine, vinegar and other beverages in bottles of silver for preventing spoilage. Silver coins were also put in milk bottles to preserve the freshness of milk [146, 148].

Years ago, the father of medicine, Hippocrates, had disclosed that antimicrobial silver had beneficial effects in healing [146]. Later silver solutions attained approval for use as an antibacterial agent, disinfectant and germicide, by the US Food and Drug Administration only in the 1920s [149, 150]. Much before antibiotics were introduced, preparations of colloidal silver were administered for ophthalmic problems, gonorrheal conjunctivitis, also as an internal medication for a variety of other diseases, like common cold and tropical spurge epilepsy [151–153]. Although silver products were frequently promoted for several medicinal benefits, the use had been even more questionable as there was the risk of some adverse side effects, e.g. argyria [154]. Together with the rising concern about the adverse effects of silver [155], the use diminished around 1940s.

Since then strategies to overcome the shortcomings started emerging [156] and applications diversifying. Surface coatings, with a thin layer of silver metal, most probably will limit even the mild toxicity of the silver toward the human body while still exercising infection control. Coating a pin with any silver-containing compound will minimize microbial colonization and pin tract infection [157] is now a confirmed hypothesis, through the *in vitro* and *in vivo* studies [143]. Silver's antimicrobial, low-toxic and hypoallergenic qualities continue to ensure its usefulness in the medical field (e.g. coating of biomedical implants, devices and sensors, creation of self-sterilizing biomaterials or as wound care especially for burn victims) where it may begin to outshine drastically now with the focused attention on the application of oligodynamic silver nanoparticles [158, 159].

1.4.4   INNOVATION OF SILVER NANOPARTICLES

Currently, the use of silver has experienced a dramatic revival, especially in its nanoparticulate form [160, 161]. Silver nanoparticles have been thoroughly tested in different fields, in science, engineering and technology, as an effective antimicrobial agent against Gram-positive as well as Gram-negative bacteria [162, 163]. Several bacteria already have become antibiotic resistant; therefore, an urgent need for developing a substitute for antibiotics arises in the foreseeable future. Silver nanoparticles appear to be attractive as they are non-toxic to humans at reduced concentration, and at the same level having broad-spectrum antimicrobial quality [164]. Silver nanoparticles can fairly resist microbial growth at even minimal concentration. Their mode of action is more effective than the bulk metal based on their enhanced surface properties. Due to several other remarkable properties and superiorly strong antimicrobial activity at the nanoscale, to date, silver nanoparticles are immensely used in a number of medical and other consumer products recently [165–168]. Hence, silver nanoparticles are at present essentially in great demand, especially by the sector of medicine and biomedical industry [169, 170].

1.4.5   PERSPECTIVE OF SILVER NANOPARTICLES

An increase in manufacture and utilization of silver nanoparticles in products have also led to an upswing in the release of silver particles into the environment at every step, starting from the raw materials to its disposal, i.e. from so-called 'the cradle to the grave.' Therefore, knowledge regarding the methods of synthesis of nanosilver particles is immensely crucial from health and environment point of view. This would allow production of ecological and human friendly silver nanoparticles along with the information of its physical, chemical and morphological characteristics. It will be an important step to focus on the evaluation of their transport, toxicity and environmental fate, along with their wide spread applications [171, 172].

1.4.6 Protocols of Synthesis

In the synthesis part of nanoparticles, main challenge is to control the physical properties like attaining identical shape, crystal structure, morphology and uniform particle size distribution, along with their desired chemical composition [173]. The shape of nanosilver particles plays an important role in its functional properties. For example, silver nanoparticles having triangular shape, have strong bactericidal action against *Escherichia coli*, the Gram-negative bacteria, as compared to that of spherical and rod shaped [174]. This is reportedly attributed due to the possible alignment of the metal atoms inside the crystal structure.

All methods of silver nanostructure synthesis basically follow common approaches with limited differences at most to the use of specific reactants and reaction conditions. Still the synthesis methods which are present now can be categorized under certain headings, such as bottom-up vs. top-down, non-green vs. green and non-conventional vs. conventional. Most usually top-down strategies involve physical methods, while in bottom-up techniques a silver salt precursor is used which is reduced from its oxidized state in the chemical reaction. Conventional synthesis methods include the use of reagents like borohydride, citrate, organic reducers, inverse micelles two-phase systems, etc., in the process of synthesis. Whereas, unconventional synthesis methods include evaporation of metal, laser ablation, catalysis, electro-condensation, etc. [171]. Green synthesis particularly represent reaction proceeded following the set of 12 principles drafted for green chemistry [175].

1.4.7 Renewable Bioresource for Silver Nanoparticle Synthesis – A Green Approach

A green approach, using organic materials or biological forms like microorganisms or plant biomass, is an alternative to physical and chemical methods for the fabrication of nanoparticles in an eco-friendly and economic way. The dark side of the green approach is uncontrolled morphological features of synthesized silver nanoparticles, as achieved by the non-green methods. The challenge is to overcome this problem.

1.5 WASTE TO VALUE CONVERSION EN ROUTE NANOBIOTECHNOLOGY

Green synthesis of nanoparticles has been achieved through several techniques [176–183] but the use of plants is preferred because it is rapid, cost-efficient, environmentally benign as well as a single-step method and can be efficiently used for large scale production [184]. Moreover, it also paves an avenue for waste to value conversion en route nanobiotecnology, through utilization of discarded plant materials, e.g. banana peels, orange peels, departed foliage, shed flowers, cast off parts, dumped bouquets, etc., without disturbing the ecology for benefactor cause. Depending upon the plants and their parts, i.e. phytomass, the difference in the constituents as well as the phytochemicals, affects the shape, size and composition of the synthesized nanoparticles [185]. The nanoparticles that are phytomass synthesized do not get agglomerated, if the particle surface is not capped with any stabilizing ligand, overcoming the problem of the requirement for the particles to be suspended into a dispersant in order to prevent cluster formation.

Therefore different morphologies of the nanoparticles, such as shape, size, texture, phase and surface properties, depend upon the method of synthesis and dispersants used [186] or lack thereof, resulting into diversity of characteristics affecting the toxicity level and the fate of the fabricated silver nanoparticles. Usually, silver nanoparticles with smaller size and in agglomeration free form possess higher antimicrobial property as compared to aggregates of nanoparticles or particles with a larger size, while these larger ones pose lesser harm to the human health and to the environment than do the smaller ones.

1.6 NANOBIOMATERIALS AND RISK ASSESSMENT: ASPECTS OF TOXICITY AND BIOCOMPATIBILITY

Nanotechnology with its concurrent applications in almost every field has enormous benefits, for the human world, but small nanoparticles can possibly accumulate in cells of the human body and may cause adverse health effects, which are at present not known to us. Risk assessment of nanomaterials mean we should thoroughly assess nanotechnology's invisible threat and be aware of the possible hazards to prevent the boon from becoming a bane, prohibiting the big consequences of the small science.

The novel properties of nanomaterials also indicate that risk assessments, which are designed for ordinary materials, may not suffice in determining the environmental and health hazards posed by nanotechnology. While there exists no documented case of any specific harm surfaced due to a nanomaterial, but there are growing evidences pointing toward unexpected risks affecting our environment and health [187, 188]. Moreover, since concentration-dependent cytotoxicity has already been demonstrated for silver-containing biomaterials, implications and regulations need to be laid down for using silver nanomaterials for biomedical and other applications with caution [189, 190].

There is a need of hour for developing proper methods, tools and techniques for the thorough characterization, detection and measuring the concentration, surface charge, chemistry and transformations of silver nanomaterials. Characterization tools like electron microscopic techniques, energy-dispersive X-ray analysis, thermal gravimetric analysis, DSC, Fourier transformed infrared spectroscopy, zeta potential size analyzer, atomic absorption spectroscopy and cytotoxicity testing, among others are helpful in the process of assessment [191]. Researchers, to address this particular issue to investigate the fate, transport and toxicity of silver nanomaterials, have adopted well-defined and documented methodology and comprehensive assessment tools like Life Cycle Assessment (LCA) together with Comprehensive Environmental Assessment (CEA) [171].

Silver nanomaterials, though having few potential risks, are outweighed with respect to the benefits to the society, through their development and deployment [192]. Above all, among the priority list of most hazardous substances and prevalent heavy metals, which are harmful for public health, silver has not been considered a threat [193].

At the heart of the introduction and application of self-sterilizing nanocomposite biomaterials in the landscape of nanomedicine [194–207], there is a strong emphasis for ensuring their complete biocompatibility. Proper investigation and comparison on several aspects in this direction is needed for bringing them into safe medical practice. They should be evaluated thoroughly on the specific basis of chemical composition, toxicological levels, physical features, morphological structure, thermal properties, mechanical characteristics and biological parameters for safely incorporating into the human body. Harmful effects are generally caused by contaminants produced by the biomaterial components, or leached out breakdown products that bring biological effects. According, to newly drafted part 18, the addition made to the ISO 10993 standards, characterization of chemical composition will potentially evaluate probable leachable substances, degradation products as well as bioavailability, while characterization of morphological structure will examine the surfaces of biomaterials. Physical and mechanical characterization, such as hardness testing, thermal analysis, etc., will address functionality and safety. Information regarding the synthesis and characteristics of the polymer also is very important together with the description of the solvents used and additives added for the preparation of any polymer nanocomposite [208–212]. The advancement of nanostructured composite materials into biomedical engineering [213] is still at its beginning and knowledge regarding the health impact with the biocompatibility factor of these biomaterials is bare essential before they can be established and standardized as the next generation biomaterials having favorable benefit-to-risk ratio.

REFERENCES

1. R. P. Feynman, Science, 254 (1991) 1300–1301.
2. R. P. Feynman, Resonance, September (2011) 890–905.
3. N. Taniguchi, On the Basic Concept of 'NanoTechnology,' Proc. Intl. Conf. Prod. Eng. Tokyo, Part II, Japan Society of Precision Engineering (1974).
4. M. Daniel and D. Astruc, Chem. Rev., 104 (2004) 293–346.
5. G. A. van Albada and J. O. Schreck, J. Chem. Educ., 64 (1987) 869–874.
6. D. J. Barber and I. C. Freestone, Archaeometry, 32 (1990) 33–45.
7. D. D. Evanoff Jr. and G. Chumanov, Chem. Phys. Chem., 6 (2005) 1221–1231.
8. M. Reibold, P. Paufler, A. A. Levin, W. Kochmann, N. Pätzke and D. C. Meyer, Nature, 444 (2006) 286.
9. C. Shao, H. Zhao and P. Wang, Nano Convergence, 9 (2022) 1–17. https://doi.org/10.1186/s40580-022-00302-0

10. M. Primožič, Z. Knez and M. Leitgeb, Nanomaterials, 11 (2021) 292. https://doi.org/10.3390/nano11020292
11. E. Inshakova, A. Inshakova and A. Goncharov, IOP Conf. Series, 971 (2020) 032031. doi:10.1088/1757-899X/971/3/032031
12. J. M. Metselaar and T. Lammers, Drug Deliv. Transl. Res., 10 (2020) 721–725. https://doi.org/10.1007/s13346-020-00740-5
13. D. Astruc, Introduction: Nanoparticles in Catalysis, Chem. Rev., 120 (2020) 461–463. http://dx.doi.org/10.1021/acs.chemrev.8b00696
14. U. Banin, N. Waiskopf, L. Hammarström, G. Boschloo, M. Freitag, E. M. J. Johansson, J. Sá, H. Tian, M. B. Johnston, L. M. Herz, R. L. Milot, M. G. Kanatzidis, W. Ke, I. Spanopoulos, K. L. Kohlstedt, G. C. Schatz, N. Lewis, T. Meyer, A. J. Nozik, M. C. Beard, F. Armstrong, C. F. Megarity, C. A. Schmuttenmaer, V. S. Batista and G. W. Brudvig, Nanotechnology, 32 (2021) 042003. https://doi.org/10.1088/1361-6528/abbce8
15. D. Manno, E. Filippo, M. Di Giulio and A. Serra, J. Non Cryst. Solids, 354 (2008) 5515–5520.
16. Y. J. Kim, Y. S. Yang, S. Ha, S. M. Cho, Y. S. Kim, H. Y. Kim, H. Yang and Y. T. Kim, Sensor. Actuat. B-Chem., 106 (2005) 189–198.
17. A. K. Rana, S. B. Rana, A. Kumari and V. Kiran, Int. J. Recent Trends in Eng., 1 (2009) 46–48.
18. F. Pacheco-Torgal and S. Jalali, Constr. Build. Mater., (2010) 1–9. doi:10.1016/j.conbuildmat.2010.07.009
19. D. E. Effiong, T. O. Uwah, E. U. Jumbo and A. E. Akpabio, Advances in Nanoparticles, 9 (2020) 1–22. https://doi.org/10.4236/anp.2020.91001
20. B. S. Hassan, G. M. N. Islam and A. N. M. A. Haque, Adv. Res. Text. Eng., 4 (2019) 1038.
21. M. Taran, M. Safaei, N. Karimi and A. Almasi, Biointerface Res. Appl. Chem., 11 (2021) 7860–7870. https://doi.org/10.33263/BRIAC111.78607870
22. A. Globus, D. Bailey, J. Han, R. Jaffe, C. Levit, R. C. Merkle and D. Srivastava, J. Br. Interplanet. Soc., 51 (1998) 145–152.
23. M. Shafique and X. Luo, Materials, 12 (2019) 2493. doi:10.3390/ma12152493
24. A. M. Ionescu, Connections QJ, 15, 2 (2016) 31–47. http://dx.doi.org/10.11610/Connections.15.2.03
25. S. Sachdeva, A. Mani, S. A. Mani, H. R. Vora, S. S. Gholap and J. K. Sodhi, IP- IJP I., 6 (2021) 6–10. https://doi.org/10.18231/j.ijpi.2021.002
26. J. Knoch, Nanoelectronics, De Gruyter Oldenbourg, Berlin and Boston, MA (2020). https://doi.org/10.1515/9783110575507
27. C. Mounica, J. Comput. Sci. Syst. Biol., 13 (2020) 328.
28. V. Ermolov, M. Heino, A. Kärkkäinen, R. Lehtiniemi, N. Nefedov, P. Pasanen, Z. Radivojevic, M. Rouvala, T. Ryhänen, E. Seppälä and M. A. Uusitalo, Significance of Nanotechnology for Future Wireless Devices and Communications, The 18th Annual IEEE International Symposium on Personal, Indoor and Mobile Radio Communications (PIMRC) (2007).
29. B. Bhushan, Introduction to Nanotechnology, In: B. Bhushan (ed.), Handbook of Nanotechnology (2nd ed.), Springer, Heidelberg (2010) 1–12.
30. K. J. Klabunde (ed.), Nanoscale Materials in Chemistry, John Wiley & Sons, Inc., New York, NY (2001).
31. G. B. Kauffman, Foundations of Chemistry, In: K. J. Klabunde and R. M. Richards (eds.), Nanoscale Materials in Chemistry (2nd ed.), 14 (2012) 183–184.
32. L. Foster, Nanotechnology Science, Innovation and Opportunity, Prentice Hall, New York, NY (2005).

33. R. Taylor, J. P. Hare, A. K. Abdul-Sada and H. W. Kroto, J. Chem. Soc. Chem. Commun., 20 (1990) 1423–1425.
34. K. E. Drexler, Engines of Creation: The Coming Era of Nanotechnology, Anchor Press, New York, NY (1986).
35. K. E. Drexler, C. Peterson and G. Pergamit, Unbounding the Future: The Nanotechnology Revolution, William Morrow and Company, New York, NY (1991).
36. A. Surendranath and M. P. Valappil, Nanomedicine: Challenges and Future Perspectives. In: D. Thangadurai, J. Sangeetha and R. Prasad (eds.), Functional Bionanomaterials. Nanotechnology in the Life Sciences, Springer, Cham (2020). https://doi.org/10.1007/978-3-030-41464-1_19
37. C. Zhang, L. Yan, X. Wang, S. Zhu, C. Chen, Z. Gu and Y. Zhao, Nano Today, 35 (2020) 101008.
38. A. Gayathri and T. Satyanarayana, Inter. J. Adv. Manag. Technol. Eng. Sci., 8 (2018) 698–705.
39. K. Kono, Polym. J., 44 (2012) 531–540.
40. N. M. Pugno, J. Phys., 18 (2006) S1971–S1990.
41. W. Ahmed, M. J. Jackson, I. ul Hassan, Nanotechnology to Nanomanufacturing, In: W. Ahmed and M. J. Jackson (eds.), Emerging Nanotechnologies for Manufacturing, Elsevier Inc., Amsterdam (2009) 1–15.
42. D. Minoli, Basic Molecular/Nanotechnology Instrumentation, Nanotechnology Applications to Telecommunications and Networking, John Wiley & Sons, Inc., Hoboken, New Jersey (2006) 308–335.
43. M. Botes and T. E. Cloete, Crit. Rev. Microbiol., 36 (2010) 68–81.
44. I. M. Chung, G. Rajakumar, T. Gomathi, S. K. Park, S. H. Kim and M. Thiruvengadam, Front. Life Sci., 10 (2017) 63–72. https://doi.org/10.1080/21553769.2017.1365775
45. P. J. J. Alvarez, C. K. Chan, M. Elimelech, N. J. Halas and D. Villagrán, Nat. Nanotechnol., 13 (2018) 634–641. https://doi.org/10.1038/s41565-018-0203-2
46. M. Usman, M. Farooq, A. Wakeel, A. Nawaz, S. A. Cheema, H. ur Rehman, I. Ashraf and M. Sanaullah, Sci. Total Environ., 721 (2020) 137778.
47. J. Altmann, Secur. Dialogue, 35 (2004) 61–79.
48. R. A. Freitas Jr., Nanomedicine, Volume I: Basic Capabilities, Landes Bioscience, Austin, TX (1999).
49. J. Panyam and V. Labhasetwar, Adv. Drug. Deliv. Rev., 55 (2003) 329–347.
50. Nanomedicine, www.nanobio-raise.org
51. X. Zhan, M. Gao, Y. Jiang, W. Zhang, W. M. Wong, Q. Yuan, H. Su, X. Kang, X. Dai, W. Zhang, J. Guo and W. Wu, Nanomedicine, 9 (2013) 305–315.
52. M. Zhang, Nanomedicine, 8 (2013) 313–316.
53. J. Lisziewicz and E. R. Tőke, Nanomedicine, 9 (2013) 28–38.
54. B. Szefler, Int. J. Nanomed., 13 (2018) 6143–6176. http://dx.doi.org/10.2147/IJN.S172907
55. L. Ye, L. Kollie, X. Liu, W. Guo, X. Ying, J. Zhu, S. Yang and M. Yu, Molecules, 26 (2021) 3252. https://doi.org/10.3390/molecules26113252
56. M. Solomon and G. G. M. D'Souza, Curr. Opin. Pediatr., 23 (2011) 215–220.
57. D. Hofmann and V. Mailander, Nanomedicine, 8 (2013) 321–323.
58. V. Wagner, A. Dullaart, A. K. Bock and A. Zweck, Nat. Biotechnol., 24 (2006) 1–9.
59. Commission Recommendation 2011/696/EU, OJ L 275 (2011).
60. F. Klaessig, M. Marrapese and S. Abe, Current Perspectives in Nanotechnology Terminology and Nomenclature, In: V. Morashov and J. Howard (eds.), Nanotechnology Standards, Springer, New York, NY (2011) 21–52.

61. F. Trotta and A. Mele, Nanomaterials: Classification and Properties, In: F. Trotta and A. Mele (eds.), Nanosponges: Synthesis and Applications (1st ed.), Wiley-VCH Verlag GmbH & Co., KGaA, Weinheim, Germany (2019) 1–26.
62. E. B. Dickerson, E. C. Dreaden, X. Huang, I. H. El-Sayed, H. Chu, S. Pushpanketh, J. F. McDonald and M. A. El-Sayed, Cancer Lett., 269 (2008) 57–66.
63. R. E. Rosensweig, J. Magn. Magn. Mater., 252 (2002) 370–374.
64. R. Bhattacharya, C. R. Patra, A. Earl, S. Wang, A. Katarya, L. Lu, J. N. Kizhakkedathu, M. J. Yaszemski, P. R. Greipp, D. Mukhopadhyay and P. Mukherjee, Nanomed. Nanotechnol. Biol. Med., 3 (2007) 224–238.
65. M. C. Urbina, S. Zinoveva, T. Miller, C. M. Sabliov, W. T. Monroe and C. S. S. R. Kumar, J. Phys. Chem. C, 112 (2008) 11102–11108.
66. C. J. Murphy, T. K. Sau, A. M. Gole, C. J. Orendorff, J. Gao, L. Gou, S. E. Hunyadi and T. J. Li, Phys. Chem. B, 109 (2005) 13857–13870.
67. M. N. Nadagouda and R. S. Varma, Green Chem., 10 (2008) 859–862.
68. R. Shukla, S. K. Nune, N. Chanda, K. Katti, S. Mekapothula, R. R. Kulkarni, W. V. Welshons, R. Kannan and K. V. Katti, Small, 4 (2008) 1425–1436.
69. M. R. Rasch, E. Rossinyol, J. L. Hueso, B. W. Goodfellow, J. Arbiol and B. A. Korgel, Nano Lett., 10 (2010) 3733–3739.
70. A. A. Pradhan, S. I. Shah and L. Pakstis, Synthesis and Characterization of Metal Nanoparticles and the Formation of Metal Polymer Nanocomposites, Mat. Res. Soc. Symp. Proc., Materials Research Society, 740 © (2003) 179–184.
71. S. Papp, R. Patakfalvi and I. Dekany, Progr. Colloid Polym. Sci., 125 (2004) 88–95.
72. P. S. Kishore, B. Viswanathan and T. K. Varadarajan, Nanoscale Res. Lett., 3 (2007) 14–20.
73. F. L. Matthews and R. D. Rawlings, Composite Materials; Engineering and Science, Chapman & Hall, London; New York (1994).
74. S. C Sharma, Composite Materials, Narosa Publishing House, New Delhi, India (2000).
75. J. Y. Rho, L. Kuhn-Spearing and P. Zioupos, Med. Eng. Phys., 20 (1998) 92–102.
76. P. Fratzl, Fibre Diffr. Rev., 210 (2002) 32–39.
77. W. T. Butler and H. Ritchie, Int. J. Dev. Biol., 39 (1995) 169–179.
78. E. Dujardin and S. Mann, Adv. Mater., 14 (2002) 775–788.
79. V. Thomas, M. Namdeo, Y. M. Mohan, S. K. Bajpai and M. Bajpai, J. Macromol. Sci. Part A, 45 (2007) 107–119.
80. F. Faupel, V. Zaporojtchenko, T. Strunskus, M. Elbahri, Adv. Eng. Mate., 12 (2010) 1177–1190.
81. O. M. Folarin, E. R. Sadiku and Arjun Maity, Int. J. Phys. Sci., 6 (2011) 4869–4882.
82. D. R. Paul and L. M. Robeson, Polymer, 49 (2008) 3187–3204.
83. E. W. Gacitua, A. A. Ballerini and J. Zhang, Ciencia y Tecnología, 7 (2005) 159–178.
84. S. K. Mazumder, Composites Manufacturing, Materials, Product and Process Engineering, CRC Taylor & Francis, Boca Raton (2002). ISBN 0-8493-0585-3.
85. M. S. Islam, W. S. Choi, Y. B. Lee and H. J. Lee, J. Mater. Chem. A, 1 (2013) 3565–3574.
86. A. Lagashetty and A. Venkataraman, Resonance, 10 (2005) 49–60.
87. B. S. Murty, P. Shankar, B. Raj, B. B. Rath and J. Murday, Tools to Characterize Nanomaterials, Textbook of Nanoscience and Nanotechnology, Springer, Berlin Heidelberg (2013) 149–175.
88. R. A. Hule and D. J. Pochan, MRS Bull, 32 (2007) 354–358.
89. E. Dujardin and S. Mann, Adv. Mater., 14 (2002) 775–788.
90. K. K. Yang, X. L. Wang and Y. Z. Wang, J. Ind. Eng. Chem., 13 (2007) 485–500.
91. B. D. Ratner, Polym. Int., 56 (2007) 1183.

92. D. Campoccia, L. Montanaro and C. R. Arciola, Biomaterials, 27 (2006) 2331–2339.

93. D. F. Williams, Biomaterials, 29 (2008) 2941–2953.

94. J. Parvizi, V. Antoci Jr., N. J. Hickok and I. M. Shapiro, Exp. Rev. Med. Dev., 4 (2007) 55.

95. A. G. Gristina, Clin. Orthop. Relat. Res., 298 (1994) 106–118.

96. J. M. Steckelberg and D. R. Osmon, Prosthetic Joint Infections, In: A. L. Bisno and F. A. Waldvogel (eds.), Infection Associated with Indwelling Medical Devices, American Society for Microbiology, Washington DC (1994) 259–901.

97. Y. H. An and R. J. Friedmann, J. Biomed. Mater. Res., 43 (1998) 338–348.

98. J. M. Schierholz and J. Beuth, J. Hosp. Infect., 49 (2001) 87–93.

99. R. T. Southwood, J. L. Rice, P. J. McDonald, P. H. Hakendorf and M. A. Rozenbilds, J. Bone Jt. Surg. Br., 67 (1985) 229–231.

100. S. Jabbouri and I. Sadovskaya, Immunol. Med. Microbiol., 59 (2010) 280–291.

101. J. A. Urban and K. L. Garvin, Curr. Treat. Options Infect. Dis., 5 (2003) 309–321.

102. W. E. Kloos and T. L. Bannerman, Clin. Microbiol. Rev., 7 (1994) 117–140.

103. A. M. Lowe, D. T. Beattie and R. L. Deresiewicz, Mol. Microbiol., 27 (1998) 967–976.

104. R. P. Tanoira, A. A. Aarnisalo, K. K. Eklund, X. Han, A. Soininen, V. M. Tiainen, J. Esteban and T. J. Kinnari, Surg. Infect., 18 (2017) 336–344. doi: 10.1089/sur.2016.263

105. S. Scialla, G. Martuscelli, F. Nappi, S. S. A. Singh, A. Iervolino, D. Larobina, L. Ambrosio and M. G. Raucci, Polymers, 13 (2021) 1556. https://doi.org/10.3390/polym13101556

106. S. Bakhshandeh and S. A.Yavari, J. Mater. Chem. B, 6 (2018) 1128–1148. https://doi.org/10.1039/C7TB02445B

107. L. Montanaro, D. Campoccia and C. R. Arciola, Int. J. Artif. Organs, 31 (2008) 771–776.

108. G. Colon, B. C. Ward and T. J. Webster, J. Biomed. Mater. Res. A, 78 (2006) 595–604.

109. Y. H. Tsuang, J. S. Sun, Y. C. Huang, C. H. Lu, W. H. Chang and C. C. Wang, Artif. Organs, 32 (2008) 167–74.

110. X. Chen and H. J. Schluesener, Toxicol. Lett., 176 (2008) 1–12.

111. S. Arora, J. Jain, J. M. Rajwade and K. M. Paknikar, Toxicol. Lett., 179 (2008) 93–100.

112. S. Saint, J. G. Elmore, S. D. Sullivan, S. S. Emerson and T. D. Koepsell, Am. J. Med., 105 (1998) 236–241.

113. U. Klueh, V. Wagner, S. Kelly, A. Johnson, J. D. Bryers, J. Biomed. Mater. Res., 53 (2000) 621–631.

114. D. P. Dowling, K. Donnelly, M. L. Mc Connell, R. Eloy and M. N. Arnaud, Thin Solid Films, 398 (2001) 602–606.

115. J. E. Gray, P. R. Norton, C. L. Marolda, M. A. Valvano and K. Griffiths, Biomaterials, 24 (2003) 2759–2765.

116. R. Saito, S. Okamura, K. Ishizu, Polymer, 34 (1993) 1189–1195.

117. V. Alt, T. Bechert, P. Steinrucke, M. Wagner, P. Seidel, E. Dingeldein, E. Domann and R. Schnettler, Biomaterials, 25 (2004) 4383–4391.

118. R. L. Williams, P. J. Doherty, D. G. Vince, G. J. Grashoff and D. F. Williams, Crit. Rev. Biocompat., 5 (1989) 221–223.

119. T. J. Berger, J. A. Spadaro, S. E. Chapin and R. O. Becher, Antimicrob. Agents Chemother., 9 (1976) 357–358.

120. A. D. Russel and W. B. Hugo, Prog. Med. Chem., 31 (1994) 351–370.

121. R. M. Slawson, M. I. Vandyke, H. Lee and J. T. Trevors, Plasmid, 27 (1992) 72–77.

122. J. A. Spardaro, S. E. Chase and D. A. Webster, J. Biomed. Mater. Res., 20 (1986) 565–577.

123. T. Gilchrist, D. M. Healy and C. Drake, Biomaterials, 12 (1991) 76–78.

124. R. Kumar and H. Munstedt, Biomaterials, 26 (2005) 2081–2088.

125. W. Lesniak, A. U. Bielinska, K. Sun, K. W. Janczak, X. Shi, J. R. Baker Jr. and L. P. Balogh, Nano Lett., 5 (2005) 2123–2130.
126. K. A. Khalil, H. Fouad, T. Elsarnagawy and F. N. Almajhdi, Int. J. Electrochem. Sci., 8 (2013) 3483–3493.
127. A. Butts, The Physical Properties of Silver, In: A. Butts and C. D. Coxe (eds.), Silver—Economics, Metallurgy, and Use, Van Nostrand Reinhold, Princeton, NJ (1967) 104–122.
128. D. R. Smith and F. R. Fickett, J. Res. Natl. Inst. Stand. Technol., 100 (1995) 119–171.
129. S. F. Etris, Silver and Silver Alloys, Kirk-Othmer Encyclopedia of Chemical Technology (4th ed.), John Wiley & Sons, Inc., New York, NY, 22 (1997) 163–179.
130. W. C. Butterman and H. E. Hilliard, Silver. Open-File Report 2004–1251. U.S. Geological Survey, Reston, Virginia (2005) 1–40.
131. C. R. Cappel, Silver Compounds, Kirk-Othmer Encyclopedia of Chemical Technology (4th ed.), John Wiley & Sons, Inc., New York, NY, 22 (1997) 179–195.
132. G. Schneider, Can. Med. Assoc. J., 131 (1984) 193–196.
133. W. L. Newton and M. F. Jones, Effectiveness of Silver Ions against Cysts of Entamoeba Histolytica, CAB Direct, 41 (1949) 1027–1034.
134. H. S. Carr, T.J. Wlodkowski and H.S. Rosenkranz, Silver Sulfadiazine: In Vitro Antibacterial Activity, Antimicrob. Agents Chemother., 4(1973) 585–587.
135. Y. Ueda, M. Miyazaki, K. Mashima, S. Takagi, S. Hara, H. Kamimura and S. Jimi, Microorganisms, 8 (2020) 1551. https://doi.org/10.3390/microorganisms8101551
136. T. Wlodkowski and H. Rosenkranz, Lancet, 29 (1973) 739–740.
137. T. Berger et al., Antimicrob. Agents Chemother., 10 (1976) 856–60.
138. R. Thurman and C. Gerba, CRC Crit. Rev. Envir. Control, 18 (1989) 295–315.
139. R. Ersek, Ann. Plast. Surg., 13 (1984) 482–487.
140. C. Hu, J. Trauma, 28 (1988) 1488–1492.
141. J. Johnson, J. Infect. Dis., 162 (1990) 1145–1150.
142. C. Collinge, G. Goll, D. Seligson and J. Easley, Orthopedics, 17 (1994) 445–448.
143. A. Masse, A. Bruno, M. Bosetti, A. Biasibetti, M. Cannas and P. Gallinaro, J. Biomed. Mater. Res. (Appl Biomater), 53 (2000) 600–604.
144. M. A. Franco-Molina, E. Mendoza-Gamboa, C. A. Sierra-Rivera, R. A Gómez-Flores, P. Zapata-Benavides, P. Castillo-Tello, J. M. Alcocer-González, D. F Miranda-Hernández, R. S Tamez-Guerra and C. Rodríguez-Padilla, J. Exp. Clin. Cancer Res., 29 (2010) 1–7.
145. C. N. Banti and S. K. Hadjikakou, Metallomics, 5 (2013) 569–596.
146. J. W. Alexander, Surg. Infect., 10 (2009) 289–292.
147. S. Medici, M. Peana, V. M. Nurchi and M. A. Zoroddu, J. Med. Chem., 62 (2019) 5923–5943. 10.1021/acs.jmedchem.8b01439
148. N. Grier, Silver and Its Compounds, In: S. S. Block (ed.), Disinfection, Sterilization and Preservation, Lea & Febiger, Philadelphia, PA (1968) 375–398.
149. FDA Regulation 21 CFR Part 310 Docket No. 96N-0144 Over-the-Counter Drug Products Containing Colloidal Silver Ingredients or Silver Salts (US Food and Drug Administration. Docket Management Branch. Rockville, Maryland. OTC Volume 100037–100039 (1973).
150. A. B. Searle, Colloids as Germicides and Disinfectants, The Use of Colloids in Health and Disease, Constable & Co., London (1920) 67–111.
151. B. G. Duhamel, Lancet., 1 (1912) 89–90.
152. A. L. Roe, Br. Med. J., 16 (1915) 104.
153. T. H. Sanderson-Wells, Lancet., 1 (1916) 258–259.
154. W. R. Hill and D. M. Pillsbury, Argyria–The Pharmacology of Silver, Williams & Wilkins, Baltimore, MD (1939).

155. M. C. Fung and D. L. Bowen, Clin. Toxicol., 34 (1996) 119–126.
156. S. Begley, Newsweek, Mar., 28 (1994) 46–51.
157. C. Collinge, G. Goll, D. Seligson and J. Easley, Orthopedics, 17 (1994) 445–448.
158. L. S. Nair and C. T. Laurencin, J. Biomed. Nanotechnol., 3 (2007) 301–316.
159. G. A. K. Reddy, J. M. Joy, T. Mitra, S. Shabnam and T. Shilpa, Int. J. Adv. Pharm., 2 (2012) 9–15.
160. M. Rai, A. Yadav and A. Gade, Biotechnol. Adv., 27 (2009) 76–83.
161. B. Nowack, H. F. Krug and M. Height, Environ. Sci. Technol., 45 (2011) 1177–1183.
162. V. K. Sharma, R. A. Yngard and Y. Lin, Adv. Colloid and Interface Sci., 145 (2009) 83–96.
163. M. Guzman, J. Dille and S. Godet, Nanomedicine, 8 (2012) 37–45.
164. S. Prabhu and E. K. Poulose, Int. Nano Lett., 2 (2012) 1–10.
165. A. Kumar, P. K. Vemula, P. M. Ajayan and G. John, Nat. Mater., 7 (2008) 236–241.
166. A. Sironmani and K. Daniel, Silver Nanoparticles – Universal Multifunctional Nanoparticles for Bio Sensing, Imaging for Diagnostics and Targeted Drug Delivery for Therapeutic Applications, In: I. Kapetanović (ed.), Drug Discovery and Development – Present and Future, InTech, Janeza Trdine 9, 51000 Rijeka, Croatia (2011) 463–488.
167. N. Rezvani, A. Sorooshzadeh and N. Farhadi, World Acad. Sci. Eng. Technol., 61 (2012) 606–611.
168. C. Ong, J. Z. Z. Lim, C. T. Ng, J. J. Li, L. Y. L.Yung, and B. H. Bay, Curr. Med. Chem., 20 (2013) 772–781.
169. K. Chaloupka, Y. Malam and A. M. Seifalian, Trends Biotechnol., 28 (2010) 580–588.
170. V. D. Lago, L. França de Oliveira, K. de Almeida Gonçalves, J. Kobarg and M. B. Cardoso, J. Mater. Chem., 21 (2011) 12267–12273.
171. A. El-Badawy, D. Feldhake and R. Venkatapathy, State of the Science Literature Review: Everything Nanosilver and More. Scientific, Technical, Research, Engineering and Modeling Support Final Report. EPA/600/R-10/084 August 2010, www.epa.gov
172. B. Reidy, A. Haase, A. Luch, K. A. Dawson and I. Lynch, Materials, 6 (2013) 2295–2350.
173. T. Tolaymat, A. El Badawy, A. Genaidy, K. Scheckel, T. Luxton and M. Suidan, Sci. Tot. Environ., 408 (2010) 999–1006.
174. S. Pal, Y. K. Tak and J. M. Song, Appl. Environ. Microbiol., 73 (2007) 1712–1720.
175. P. T. Anastas and J. C. Warner, Green Chem., Oxford University Press Inc, New York, NY (1998).
176. P. Raveendran, J. Fu and S. L. Wallen, J. Am. Chem. Soc., 125 (2003) 13940–13941.
177. S. S. Shankar, A. Rai, B. Ankamwar, A. Singh, A. Ahmad and M. Sastry, Nat. Mater., 3 (2004) 482–488.
178. P. Raveendran, J. Fu and S. L. Wallen, Green Chem., 8 (2006) 34–38.
179. G. Zhang, B. Keita, A. Dolbecq, P. Mialane, F. Sécheresse, F. Miserques and L. Nadjo, Chem. Mater., 19 (2007) 5821–5823.
180. M. Chandra and P. K. Das, Int. J. Green Nanotechnol., 1 (2009) P10–P25.
181. Z. Khan, J. I. Hussain, S. Kumar, A. A. Hashmi and M. A. Malik, J. Biomater. Nanobiotechnol., 2 (2011) 390–399.
182. G. A. Bhaduri, R. Little, R. B. Khomane, S. U. Lokhande, B. D. Kulkarni, B. G. Mendis and L. Šiller, J. Photochem. Photobiol. A, 258 (2013) 1–9.
183. B. Krishnaveni and P. Priya, Int. J. Chem. Stud., 1 (2014) 10–20.
184. R. P. S. Chauhan, C. Gupta and D. Prakash, Int. J. Bioassays, 1 (2012) 1–5.

185. P. Dwivedi, S. S. Narvi and R. P. Tewari, Ind. Crops Prod., 54 (2014) 22–31.
186. P. Dwivedi, D. Tiwary, P. K. Mishra and J. P. Chakraborty, Adv. Sci. Eng. Med., 12 (2020) 548–555.
187. A. D. Maynard, R. J. Aitken, T. Butz, V. Colvin, K. Donaldson, G. Oberdörster, M. A. Philbert, J. Ryan, A. Seaton, V. Stone, S. S. Tinkle, L. Tran, N. J. Walker, and D. B. Warheit, Safe Handling of Nanotechnology, Nature, 444 (2006) 267–269.
188. G. Oberdorster, V. Stone and K. Donaldson, Nanotoxicology, 1 (2007) 2–25.
189. L. Braydich-Stolle, S. Hussain, J. J. Schlager and M. C. Hofmann, Toxicol. Sci., 88 (2005) 412–419.
190. M. E. Samberg, E. G. Loboa, S. J. Oldenburg, and N. A. Monteiro-Riviere, Nanomedicine, 7 (2012) 1197–1209.
191. Final Report dated 07/15/2010 xiii State of the Science – Everything Nanosilver and More, (2010).
192. P. Dwivedi, S. S. Narvi and R. P. Tewari, Silver Nanoparticles and Nanocomposites, Encycl. Biomed. Polym. Polym. Biomater., 10 (2015) 7275–7285.
193. J. J. Harrison, H. Ceri and R. J. Turner, Nat. Rev. Microbiol., 5 (2007) 928–938.
194. P. Dwivedi, S. S. Narvi and R. P. Tewari, International Conference on Nanoscience, Technology and Societal Implications, NSTSI11(2011).
195. P. Dwivedi, S. S. Narvi and R. P. Tewari, Int. J. Green Nanotechnol., 4 (2012) 248–261.
196. P. Dwivedi, S. S. Narvi and R. P. Tewari, Adv. Mater. Res., 585 (2012) 144–148.
197. P. Dwivedi, S. S. Narvi and R. P. Tewari, Int. J. Biomed. Nanosci. Nanotechnol., 2 (2012) 187–206.
198. P. Dwivedi, S. S. Narvi and R. P. Tewari, J. Chin. Med. Res. and Develop., 1 (2012) 23–27.
199. P. Dwivedi, S. S. Narvi and R. P. Tewari, Int. J. Adv. Eng., Sci. and Technol., 2 (2012) 236–243.
200. P. Dwivedi, S. S. Narvi and R. P. Tewari, Int. J. Eng. Res. Appl., 2 (2012) 1490–1495.
201. P. Dwivedi, S. S. Narvi and R. P. Tewari, Int. J. Sci. and Res. Publ., 2, (2012) 1–5.
202. P. Dwivedi, S. S. Narvi and R. P. Tewari, Annu. Res. Rev. Biol., 4 (2014) 1059–1069.
203. P. Dwivedi, S. S. Narvi and R. P. Tewari, Adv. Sci. Eng. Med., 6 (2014) 1–9.
204. P. Dwivedi, S. S. Narvi and R. P. Tewari, Nano LIFE, 5 (2015) 1540006.
205. P. Dwivedi, D. Tiwary, S. S. Narvi, R. P. Tewari and K. P. Shukla, Lett. Appl. NanoBioScience, 9 (2020) 1485–1493.
206. P. Dwivedi, D. Tiwary, P. K. Mishra and J. P. Chakraborty, Nano-Struct. Nano-Objects, 22 (2020) 100485, 1–7.
207. P. Dwivedi, D. Tiwary, P. K. Mishra, S. S. Narvi and R. P. Tewari, Inorg. Chem. Commun., 126 (2021) 108479, 1–12.
208. Biological Evaluation of Medical Devices, ISO 10993, parts 1–13 (Geneva: International Organization for Standardization [ISO], various dates).
209. D. E. Albert and R. F. Wallin, Medical Device Diagnostic Ind., 20 (1998) 96–99.
210. J. P. Sibilia, Molecular Spectroscopy, In: J. P. Sibilia (ed.), A Guide to Materials Characterization and Chemical Analysis, VCH Publishers, New York, NY (1988) 13–40.
211. M. Bosetti, A. Masse, E. Tobin and M. Cannas, Biomaterials, 23 (2002) 887–892.
212. D. E. Albert, The Growing Importance of Materials Characterization in Biocompatibility Testing, Reprinted from Medical Device and Diagnostic Industry (March 2002), Copyright © 2002 Canon Communications: pp 1–8.
213. P. Dwivedi, S. S. Narvi and R. P. Tewari, J. Appl. Biomater. Funct. Mater., 11 (2013) 129–142.

2 Enumerate Investigations

2.1 BIOMASS FOR NANOPARTICLES: A REVIEW

Since ancient time immemorial, plants have been the potential source of medicines with innumerable curative properties [1, 2]. Down the ages, therefore and also because of their biomass diversity, plants have been particularly responsible for flourishing prosperity in varied arrays of life. Most recently plant extract mediated synthesis has become more popular [3–15], evolving into an important green branch of nanotechnology [16–21]. Albeit, there are a vast number of methods of synthesis for nanostructured materials present in the literature [22–40], during the past decade the emerging trend of metal nanoparticle synthesis using different parts of plants, i.e. the plant materials (plant biomass), has become extremely popular due to the several positive advantages it holds even over other elaborate time-consuming processes of biosynthesis [41–57]. Among various nanoparticles of metals and metal oxides, silver nanoparticles have attracted conspicuously tremendous interest [58–64] and considerable attention focuses on it for the many essential properties and of its vast potential applications in diverse fields, chiefly in medicine [65–79].

Hitherto, a consistent number of myriad reports represent the biogenesis of silver nanoparticles successfully mediated by plant biomass (phytomass), wherein the phytomass constituents act as the reducing agents as well as the stabilizing ligands. Particularly, there is extensive exploration of leaf extracts, such as *Pelargonium graveolens* [80], *Azadirachta indica* [81], *Aloe vera* [82], *Cinnamomum camphora* [83], *Diopyros kaki* [84], *Datura metel* [85], *Eucalyptus hybrid* [86], *Glycine max* [87, 88], *Hibiscus rosa sinensis* [89], *Euphorbia hirta* [90], *Acalypha indica* [91], *Cassia fistula* [92], *Pongamia pinnata* [93], *Coriandrum Sativum* [94], *Cycas* [95], *Syzygium cumini* [96], *Mangnifera indica* [97], *Zingiber officinale* [98], *Catharanthus roseus* [99], *Nicotiana tobaccum* [100], *Vitex negundo* [101], *Ocimum sanctum* [102], *Spinacia oleracea* [103], *Lactuca sativa* [103], *Murraya koenigii* [104], *Cleome viscose* [105], *Cassia auriculata* [106, 107], *Piper betle* [108], *Tecoma stans* [109], *Cardiospermum halicacabum* [110], *Tephrosia purpurea* [111], *Moringa oleifera* [112], *Andrographis paniculata* [113], *Portulaca oleracea* [114], *Ricinus communis* [115], *Paederia foetida* [116], *Artemisia nilagirica* [117], *Pterocarpus santalinus* [118], *Olax scandens* [119], *Ficus elastica* [120], *Punica granatum* [121], *Ziziphus mauritiana* [122], *Clinacanthus nutans* [123], *Eichhornia crassipes* [124], *Colocasia esculenta* [124], *Camellia sinensis* [125], etc., for synthesizing silver nanoparticles.

DOI: 10.1201/9781003217343-2

Some of the examples in which broths of parts other than the leaves have been used for synthesis are *Medicago sativa* [126], *Emblica officinalis* [127], *Carica papaya* [128, 129], *Cinnamon zeylanicum* [130], *Jatropha curcas* [131], *Syzygium aromaticum* [132, 133], *Syzygium cumini* [96], *Trianthema decandra* [134], *Allium cepa* [135], *Capsicum annuum* [136], *Citrus sinensis* [137], *Anthoceros* [138], *Shorea tumbuggaia* [139], *Calotropis procera* [140], *Cassia Auriculata* [141], *Allium sativum* [142], *Elettaria cardamomom* [143], *Solanum xanthocarpum* [144], *Dioscorea oppositifolia* [145], *Punica granatum* [146], *Rosa damascene* [146], *Millingtonia hortensis* [147], *Morinda citrifolia* [148], *Citrus limon* [149], *Citrus limetta* [149], *Rumex hymenosepalus* [150], *Portulaca oleracea* [151], *Arnebia nobilis* [152], kiwifruit [153], *Tecoma stans* [154], *Phyllanthus emblica* [155], etc. While use of grasses and weeds for this purpose is only scarce, such as Parthenium [156], *Desmodium triflorum* [157], *Panicum virgatum* [158], *Eichornia crassipes* [159], *Padina tetrastromatica* [160], water hyacinth [161], *Pulicaria glutinosa* [162], *Ipomoea carnea* [163], lemongrass [164], etc.

2.2   METAL/POLYMER NANOCOMPOSITES: AN INSIGHT

With a look into the domain of metal/polymer nanocomposites we have some of the most comprehended ones, e.g. Silver/polymer nanocomposites and palladium/polymer nanocomposites. Synthesis of hybrid inorganic/organic films in a single-step process has been investigated by Compton et al. [165]. Lee et al. reported both catalytic as well as inhibitory properties of Pd in their palladium/polymer nanocomposites [166]. Copper/polymer nanocomposites is the group of Cu-based materials which is an important class of metal containing nanocomposites. Novel *in situ* synthesis method for preparing copper/polymer nanocomposites is reported by Mallick et al. [167]. Iron/polymer nanocomposites is also an important class of materials. A facile approach for fabricating a robust vinyl ester resin composite, reinforced with iron nanoparticles, is reported by Guo et al. [168].

Aluminum/polymer nanocomposites– Huang et al. studied aluminum/polyethylene (Al/PE) nanocomposites. The authors, in their publication, in 2008, described the preparation of Al/polymer nanocomposites in the concentration range of 1–48 wt% Al particles having 102-nm diameter, dispersed in a PE matrix [169]. Gold/polymer nanocomposites– the preparation of regular and mechanically strong electric components based on Au/conducting polymer nanomaterials is investigated by Zotti et al. [170].

Nickel/polymer nanocomposites – Ghose et al. presented in their work the ways to increase the thermal conductivity of a commercially available ethylene vinyl acetate copolymer (Elvax™ 260) by compounding the copolymer with carbon-based nanofillers and including nanostrands of Ni (400–800 nm) [171]. Platinum/polymer nanocomposites – polyaniline (PANI) is a commonly used matrix for building polymer nanocomposites with Pt, exemplified by Nyczyk et al. [172]. Zinc/polymer nanocomposites – was discussed by Yang et al. [173], while magnesium/polymernanocomposite was discussed by Jeon et al. [174] for their respective applications. Cobalt/polymer nanocomposites – Karpacheva et al. prepared and evaluated the properties of homogeneously distributed Co nanoparticles in polydiphenylamine polymer matrix [175].

2.3   OVERVIEW OF SILVER/POLYMER NANOCOMPOSITES

A good number of literatures are available for this highly coveted nanomaterial and to account them all is beyond the limitations of this compilation. To recall some of them, we have nonspherical silver nanoparticles embedded in a thin plasma-polymer film matrix [176], silver/polyacrylamide (Ag/PAM) nanocomposites and silver-polymethylmethacrylate (Ag/PMMA) nanocomposites with homogeneously dispersed silver nanoparticles in polymer matrices were prepared by a new technique of ultraviolet irradiation at room temperature [177]. Silver/poly(2-hydroxyalkyl methacrylate) was also found to be a stable nanomaterial [178].

The metal-polymer interaction effects, on the structure of silver particulate films deposited on three different softened polymer substrates, namely polystyrene (PS), the poly(2-vinylpyridine) (P2VP) and poly(4-vinylpyridine) (P4VP), by vacuum evaporation technique, were also studied [179]. Formation of silver/poly(vinylpyrrolidone) (PVP) nanocomposite was efficiently performed through the reduction of Ag^+ ions with N,N-dimethyl formamide both at reflux and under microwave irradiation [180]. Dispersion polymerization induced by γ-ray irradiation was exploited to fabricate silver/polystyrene nanocomposite microspheres [181]; polyurethane-co-polystyrene films containing silver nanoparticles could also be synthesized [182]. Characteristics of silver/poly(vinylalcohol) (PVA) nanocomposites prepared by mixing a nanosilver-colloidal solution with an aqueous PVA solution in appropriate ratios was studied [183]. Silver/methoxypolyethyleneglycol (MPEG) by ultraviolet irradiation technique [184], silver/PVP by γ-irradiation [185], silver/heparin [186] and silver/chitosan by reduction method [187] were developed. Silver/poly(vinyl acetate) nanocomposite microspheres (Ag/PVAc), having high molecular weight (HMW), which are promising precursors of materials with radio-opacity, were prepared through suspension polymerization method [188].

Nanocomposite films having silver nanoparticles dispersed in a polymer matrix of Teflon AF, poly(methylmethacrylate) (PMMA) and Nylon 6 were developed by vapor phase co-deposition technique in high vacuum [189]. Silver/PMMA-PET nanocomposite structure formed by electron beam evaporation [190] and silver/poly(acrylamide-co-acrylic acid) hydrogel nanocomposite [191] have been investigated. Silver nanoparticle dispersed polyrhodanine nanofibers were prepared by chemical oxidation polymerization [192]. Study was reported for the preparation of silver nanoparticles existing with uniform distribution within the semi-interpenetrating hydrogel networks (SIHNs) on cross-linked poly(acrylamide) synthesized through an optimized rapid redox solution polymerization with N,N'-methylenebisacrylamide (MBA) in the presence of three varied carbohydrate polymers, which were starch (SR), carboxy methyl cellulose (CMC) and gum acacia (GA) [193]. The solution method was adopted to prepare the Ag/Poly(vinylidene fluoride), i.e. (Ag/PVDF) nanocomposite [194] and reduction

method for Ag/nanocomposites with polysaccharides, such as glycosaminoglycans (GAGs) [195], during the same year. Thereafter, the synthesis of Ag nanoparticles in polyethylene terephthalate (PET) matrix was reported using atom beam
co-sputtering [196], Ag/polyacrylonitrile (PAN) [197] and Ag/poly(ethylene
co-propylene) (EPR) [198] nanocomposites.

Papers presented the synthesis and properties of Ag/chitosan/gelatin [199],
Ag/chitosan/polyethylene glycol (PEG) [200], Ag/chitosan-o-methoxy polyethylene glycol [201], Ag/polyimide [202], Ag/poly (diallyldimethylammonium chloride) [203] and Ag/poly(silylene-*co*-silyne)s [204] nanocomposites. Ag/Polypyrrole
(Ppy) and Ag/Carboxymethyl cellulose (CMC) were also synthesized to put silver
nanoparticles into some practical use [205]. Nanocomposites with much utility,
such as Ag/polyaniline (PANI) [206], Ag/chitosan/PVA [207] and poly(vinyl
butyral-*co*-vinyl alcohol-*co*-vinyl acetate) (PVVV) [208], were also developed.
Possible ways to synthesize poly(1-vinyl-1,2,4-triazole) [209], Ag/(polyethylene
glycol 600 diacrylate) [210], Ag/poly vinyl pyrrolidone (PVP)/PVA blend [211]
and Ag/SiO$_2$ [212] nanocomposite materials have been investigated.

2.3.1 SILVER/CHITOSAN NANOCOMPOSITE

Today, nanocomposite materials that explore the properties of silver at the nano
level for biomedical purposes are of increasing interest and may be present in various commercial products. Silver/chitosan bionanocomposite is one among such
holding high-fertile ground in the sector of biomedical engineering [213–216]. That
is why for the silver bionanocomposite construct designed to utilize the properties
of silver at the nano range, a lot of effort has been aimed at developing efficient
and reproducible routes for the fabrication of silver nanoparticles and techniques
of fabrication for the bionanocomposite thereafter. Myriads of potential uses for
this novel nanomaterial, formed from the combination of silver and chitosan, are
being extensively investigated, involving ranges from antimicrobial therapies to
tissue engineering, because of their biocompatible properties and excellent bio-
activities [217–222].

A variety of metal/chitosan nanocomposites were developed, including silver in aqueous solutions. Silver nanoparticles were synthesized by reduction of
the silver salt with sodium borohydride (NaBH$_4$) in the presence of chitosan.
Nanoparticles adsorbed onto the surface of chitosan molecules formed the silver/
chitosan nanocomposite [187]. Chitosan-based nanocomposite film was also prepared using solvent casting method by incorporation with silver nanoparticles
[223]. Silver/chitosan nanocomposites were also synthesized by the use of basic
chitosan suspension as a stabilizing agent and as a reductant in the absence of
any other chemical, then silver nanoparticles formed through one-pot synthesis
method were attached to the polymer [224, 225]. In other case, nanosilver and chitosan combination for wound dressing purpose was fabricated using a nanometer
scale and self-assembly technology [226]. A facile approach to prepare silver/
chitosan films was done by photochemical reduction method of silver ions in an
acidic solution of chitosan and AgNO$_3$ [227]. Silver/chitosan bionanocomposites

were synthesized using the reducing agent D-glucose [228] with a reduction process conducted applying γ-radiation [229].

Silver/chitosan nanocomposites were produced in series with nanoparticles of different sizes using different molecular weight (MW) grades of chitosan through aqueous chemical reduction route [230]. Nanocomposite consisting of silver nanoparticles dispersed within chitosan matrix was well developed through crosslinking technique [231]. Chemical methods were used for synthesizing silver/chitosan nanocomposite with sodium sulfate [232] and trisodium citrate as reducing agents [233]. Silver/chitosan nanocomposite film based on silver nanoparticle biosynthesis by *Bacillus subtilis* has been achieved which is also noteworthy [234]. Henceforth, several other notable green routes and biosynthetic pathways have been adopted for achieving the same and case studies explored [235–237].

2.4   RESEARCH IN DEMAND AND BIOMEDICAL APPLICATIONS

Phytomass, which our group had chosen and were overlooked till then for the phytofabrication of nanoparticles [238–240], are:

- *Elaeocarpus ganitrus* (Rudraksha) – Seed
- *Terminalia arjuna* (Arjun) – Bark, Leaves, Fruits
- *Pseudotsuga menziesii* (Christmas tree) – Leaves
- *Prosopis spicigera* (Shami) – Leaves
- *Ficus religiosa* (Peepal) – Leaves
- *Curcuma longa* (Turmeric) – Rhizome

Silver/polymer nanocomposites, which were developed and characterized as well as reported by us with originality [241–250], also still stand novel during this compilation, are:

- Silver/chitosan bionanocomposite via phytofabrication route
- Silver/chitosan-g-polyacrylamide hydrogel nanocomposite
- Silver/chitosan/poly vinyl chloride blend nanocomposite

2.5   CONCERNED OBJECTIVES

This research is basically concerned with the design of fast, economic and safe nanotechnology to create a new pathway to connect nanostructure effect to macrostructure performance, which may take roots in a variety of diverse healthcare applications in the sector of nanomedicine. This work is centered round the current efforts and key research challenges in the development of self-sterilizing silver/polymer nanocomposite biomaterials for potent application in the arena of biomedical engineering, opposing bacterial colonization causing nosocomial infection. Investigation of the possible effectiveness of such developed biomaterials, for improving the 'quality of life' of patients globally.

Biomaterials certainly play a crucial role in modern medicine for restoring function to improve health and give vigor to life. Thus, biomaterials associated infections (BAI), occurring in approximately 0.5–6% of all cases on average, lay a serious complication and a cause of growing concern with the increase in use of biomedical devices and implants. In fact the biomaterial is possibly prone and most susceptible to infection at the time of incorporation into the body (periop-erative contamination) due to contaminated hand or hospital environment. The consequences resulting due to an infected biomaterial are much more devastat-ing as well as significant regarding the morbidity/mortality of patient and health system expenses. BAI is tough to treat, as the mode of biofilm growth protects the pathogen against the host antibodies and antibiotic treatment. From this perspec-tive, the biggest challenge therefore is the development of self-sterilizing surfaces capable of resisting infection which is at an initial stage of budding.

The future demand for research aims at increasing the antibacterial efficacy together with the biocompatible properties of materials used for biomedical applications. In the race prevailing for the surface between tissue integration and microbial adhesion, there exists an urgent universal call to ameliorate the pre-vailing face of the existing biomaterials. Therefore, our chief objective is to put forward a driving force to bring forth a winning strategy to minimize the risk of failure, offering the possibility of long lasting implants and other biomaterials, thereby reducing the number of revision operations and combat the threat of BAI or in true sense the health-care related infections which is a cause of global 'qual-ity of life' concern.

This can be achieved by the combination of few active strategies, or by cou-pling some active and passive approaches. Thought for this purpose is the need to prepare novel polymer/nanocomposites coming up from varied constructions, compositions and having different chemistries. Primarily by varying two param-eters: first, by optimizing the interaction between the polymer matrix and the dis-persed filler nanoparticles preventing their intensive agglomeration, and second, by varying the physico-chemical characteristics of the nanoscale fillers through processing technologies.

Preparation of polymer matrix nanocomposites exploring nanosilver fillers is a chosen area of research with keen interest for the foremost reason to prevent the paradigm of BAI employing a facile and pragmatic route. Thereby impart (1) surface modification of medical implants, prostheses and devices through novel coatings to inhibit bacterial adhesion and enhance implant integration, and (2) development of novel polymer mixes and blends to improve the stability of nanoscale silver fillers for the innovation of self-sterilizing nanocomposite bio-compatible biomaterials.

Mother-nature is the master chemist holding incredible talent and bearing bewildering innumerable natural reagents in her lap. Along with the grace of biodiversity which mother nature has bestowed globally, our laterally harbored main motto is to blend herbalism with nanotechnology. The intended purpose behind this is for developing green, human and eco-friendly silver nanopar-ticles of immense economic value, great critical importance and having huge applicability, especially as well as crucially in the biomedical landscape. Extracts from a wide range of different plant materials from the botanical diversity were used to mediate the fabrication of silver nanoparticles via extracellular biosynthesis (termed as phytofabrication). Phytofabrication opted to include the beneficiaries of the plants, while following the (1, 3, 4, 5, 6, 7, 8, 10 and 12) principles of 'Green Chemistry' [251].

REFERENCES

1. L. Hoareau and E. J. DaSilva, Electron. J. Biotechnol., 2 (1999) 56–70.
2. R. Verpoorte, H. K. Kim and Y. H. Choi, Plantsas Source for Medicines: New Perspectives, In: R. J. Bogers, L. E. Craker and D. Lange (eds.), Medicinal and Aromatic Plants, Springer, Netherlands (2006) 261–273.
3. M. Rai, A. Yadav and A. Gade, Crit. Rev. Biotechnol., 28 (2008) 277–284.
4. V. Kumar and S. K. Yadav, J. Chem. Technol. Biotechnol., 84 (2008) 151–157.
5. P. Rajasekharreddy, P. U. Rani and B. Sreedhar, J. Nanopart. Res., 12 (2010) 1711–1721.
6. R. Iravani, Green Chem., 13 (2011) 2638–2650.
7. L. Marchiol, Ital. J.Agron., 7 (2012) 274–282.
8. H. A. Salam, P. Rajiv, M. Kamaraj, P. Jagadeeswaran, S. Gunalan and R. Sivaraj, I. Res. J. Biological Sci., 1 (2012) 85–90.
9. Y. Zhang, G. Gao, Q. Qian and D. Cui, Nanoscale Res. Lett., 7 (2012) 475–482.
10. P. P. Gan and S. F. Y. Li, Rev. Environ. Sci. Bio/Technol., 11 (2012) 169–206.
11. B. Zheng, T. Kong, X. Jing, T. Odoom-Wubah, X. Li, D. Sun, F. Lu, Y. Zheng, J. Huang and Q. Li, J. Colloid Interface Sci., 396 (2013) 138–145.
12. A. K. Mittal, Y. Chisti and U. C. Banerjee, Biotechnol. Adv., 31 (2013) 346–356.
13. M. S. Akhtar, J. Panwar and Y. S. Yun, ACS Sustainable Chem. Eng., (2013) 12 pages A–L.
14. P. Logeswari, S. Silambarasan and J. Abraham, Scientia Iranica F., 20 (2013) 1049–1054.
15. S. K. R. Namasivayam, B. B. Christo, R. S. A. Bharani, S. M. K. Arasu, K. A. M. Kumar and K. Deepaknano, Int. J. Pharm. Pharm. Sci., 6 (2014) 376–379.
16. A. A. Lawrence and J. T. J. Prakash, Int. J. Sci. Res. Rev., 8 (2019) 267–287.
17. M. Nasrollahzadeh, M. Atarod, M. Sajjadi, S. M. Sajadi and Z. Issaabadi, Interface Sci. Technol., 28 (2019) 199–322.
18. S. Hameed, S. A. Shah, J. Iqbal, M. Numan, W. Muhammad, M. Junaid, S. Shah, R. Khurshid and F. Umer, Bioinspired, Biomim. Nanobiomaterials, 9 (2020) 95–102.
19. Y. Bao, J. He, K. Song, J. Guo, X. Zhou and S. Liu, J. Chem., Article ID 6562687 (2021) 1–14.
20. J. A. Hernández-Díaz, J. J. Garza-García, A. Zamudio-Ojeda, J. M. León-Morales, J. C. López-Velázquez and S. García-Morales, J. Sci. Food Agric., 101 (2021) 1270–1287.
21. M. Rafique, M. Sohaib, R. Tahir, M. B. Tahir and M. Rizwan, J. Nanosci. Nanotechnol., 21 (2021) 3573–3579.
22. M. Yamamoto and M. Nakamoto, J. Mater. Chem., 13 (2003) 2064–2065.
23. B. Wiley, T. Herricks, Y. Sun and Y. Xia, Nano Lett., 4 (2004) 1733–1739.
24. D. D. Evanoff Jr. and G. Chumanov, Chem. Phys. Chem., 6 (2005) 1221–1231.
25. M. Starowicz, B. Stypula and J. Banas, Electrochem. Commun., 8 (2006) 227–230.
26. S. Navaladian, B. Viswanathan, R. P. Viswanath and T. K. Varadarajan, Nanoscale. Res. Lett, 2 (2007) 44–48.

27. S. D. Solomon, M. Bahadory, A. V. Jeyarajasingam, S. A. Rutkowsky and C. Boritz, J. Chem. Edu., 84 (2007) 322–325.
28. K. J. Sreeram, M. Nidhin and B. U. Nair, Bull. Mater. Sci., 31 (2008) 937–942.
29. M. G. Guzmán, J. Dille and S. Godet, Int. J. Chem. Biol. Eng., 2 (2009) 104–111.
30. K. M. M. Abou El-Nour, A. Eftaiha, A. Al-Warthan and R. A. A. Ammar, Arab. J. Chem., 3 (2010) 135–140.
31. J. I. Hussain, S. Kumar, A. A. Hashmi and Z. Khan, Adv. Mat. Lett., 2 (2011) 188–194.
32. G. Zhou and W. Wang, Orient. J. Chem., 28 (2012) 651–655.
33. B. Baruah, G. J. Gabriel, M. J. Akbashev and M. E. Booher, Langmuir, 29 (2013) 4225–4234.
34. S. I. Hong, A. Duarte, G. A. Gonzalez and N. S. Kim, J. Electron. Packag., 135 (2013) 1–5.
35. S. K. Balavandy, K. Shameli, D. R. B. A. Biak and Z. Z. Abidin, Chem. Cent. J., 8 (2014) 1–10.
36. S. Neretina, R. A. Hughes, K. D. Gilroy and M. Hajfathalian, Acc. Chem. Res., 49 (2016) 2243–2250.
37. F. S. Al-Mubaddel, S. Haider, W. A. Al-Masry, Y. Al-Zeghayer, M. Imran, A. Haider and Z. Ullah, Arab. J. Chem., 10 (Supplement 1) (2017) S376–S388.
38. S. F. Soares, T. Fernandes, A. L. Daniel-da-Silva and T. Trindade, Proc. R. Soc. A, 475, 20180677 (2019) 1–33.
39. N. M. Noah, J. Nanomater., Article ID 8855321 (2020) 1–20.
40. M. Shepida, O. Kuntyi, M. Sozanskyi and Y. Sukhatskiy, Adv. Mater. Sci. Eng., Article ID 7754523 (2021) 1–9.
41. P. Mukherjee, S. Senapati, D. Mandal, A. Ahmad, M. I. Khan, R. Kumar and M. Sastry, Chem. Biochem., 5 (2002) 461–463.
42. M. Kowshik, S. Ashtaputre, S. Kharrazi, W. Vogel, V. W. Urban, S. K. Kulkarni and K. M. Paknikar, Nanotechnology, 14 (2003) 95–100.
43. N. Duran, P. D. Marcato, O. L. Alves, G. I. De-Souza and E. Esposito, J. Nanobiotechnol., 3 (2005) 8.
44. I. Willner, R. Baron and B. Willner, J. Adv. Mater., 18 (2006) 1109–1120.
45. P. Mohanpuria, N. K. Rana and S. K. Yadav, J. Nano Res., 10 (2008) 507–527.
46. J. Kasthuri, S. Veerapandian and N. Rajendiran, Colloids Surf. B, 68 (2009) 55–60.
47. J. Sarkar, D. Chattopadhyay, S. Patra, S. S. Deo, S. Sinha, M. Ghosh, A. Mukherjee and K. Acharya, Dig. J. Nanomater. Biostruct., 6 (2011) 563–573.
48. K. B. Narayanan and N. Sakthivel, Adv. Colloid Interfac., 169 (2011) 59–79.
49. A. A. Rahuman, C. Jayaseelan, L. Karthik, S. Marimuthu, S. Kumar, et al. Mater. Lett., 81 (2012) 69–72.
50. R. Lu, D. Yang, D. Cui, Z. Wang and L. Guo, Int. J. Nanomed., 7 (2012) 2101–2107.
51. H. Haiza, A. Azizan, A. H. Mohidin and D. S. C. Halin, Nano Hybrids, 4 (2013) 87–98.
52. R. P. Metuku, S. Pabba, S. Burra, S. V. S. S. S. L. H. Bindu, K. Gudikandula and M. A. S. Charya, Biotechnology, 4 (2014) 227–234.
53. I. W. S. Lin, C. N. Lok and C. M. Che, Chem. Sci., 5 (2014) 3144–3150.
54. S. Ahmad, S. Munir, N. Zeb, A. Ullah, B. Khan, J. Ali, M. Bilal, M. Omer, M. Alamzeb, S. M. Salman and S. Ali, Int. J. Nanomed., 14 (2019) 5087–5107.
55. A. M. El Shafey, Green Process. Synth., 9 (2020) 304–339.
56. M. A. Ali, T. Ahmed, W. Wu, A. Hossain, R. Hafeez, M. M. Islam Masum, Y. Wang, Q. An, G. Sun and B. Li, Nanomaterials, 10, 1146 (2020) 1–24.
57. H. Bahrulolum, S. Nooraei, N. Javanshir, H. Tarrahimofrad, V. S. Mirbagheri, A. J. Easton and G. Ahmadian, J. Nanobiotechnol., 19, 86 (2021) 1–26.

58. I. M. Chung, I. Park, K. Seung-Hyun, M. Thiruvengadam and G. Rajakumar, Nanoscale Res. Lett., 11 (2016) 1–14.
59. S. Jain and M. S. Mehata, Sci. Rep.,7, 15867 (2017) 1–13.
60. N. S. Jalani, S. Zati-Hanani, Y. P. Teoh and R. Abdullah, Mater. Sci. Forum, 917 (2018) 145–151.
61. R. Balachandar, P. Gurumoorthy, N. Karmegam, H. Barabadi, R. Subbaiya, K. Anand, P. Boomi and M. Saravanan, J. Clust. Sci., 30 (2019) 1481–1488.
62. A. Sharma, S. Thomas, A. M. Mathewffi and A. K. Agarwal, Mapana J. Sci., 18 (2019) 37–44.
63. S. Kaabipour and S. Hemmati, Beilstein J. Nanotechnol., 12 (2021) 102–136.
64. S. K. Chandraker, M. K. Ghosh, M. Lal and R. Shukla, Nano Express, 2, 022008 (2021) 1–22.
65. I. Sondi and B. S. Sondi, J. Colloid Interface Sci., 275 (2004) 177–182.
66. J. R. Morones, J. L. Elechiguerra, A. Camacho, K. Holt, J. B. Kouri, J. T. Ramírez, et al., Nanotechnology, 16 (2005) 2346–2353.
67. J. S. Kim, E. Kuk, K. N. Yu, J. H. Kim, S. J. Park, H. J. Lee, et al., Nanomedicine, 3 (2007) 95–101.
68. J. Y. Song and B. S. Kim, Bioprocess Biosyst. Eng., 32 (2009) 79–84.
69. A. Kaler, N. Patel and U. C. Banerjee, Curr. Res. Inform. Pharm. Sci., 11 (2010) 68–71.
70. M. Gilaki, J. Biol. Sci., 10 (2010) 465–467.
71. M. Forough and K. Farhadi, Turkish J. Eng. Env. Sci., 34 (2010) 281–287.
72. C. M. Jones and E. M. V. Hoek, J. Nanopart. Res., 12 (2010) 1531–1551.
73. N. Savithramma, M. L. Rao, K. Rukmini and P. Swarnalathadevi, Int. J. Chem. Tech. Res., 3 (2011) 1394–1402.
74. A. K. Mondal, S. Mondal, S. Samanta and S. Mallick, Adv. Biores., 2 (2011) 122–133.
75. K. Kathiresan, N. M. Alikunhi and A. Nabikhan, Int. J. Biomed. Nanosci. Nanotechnol., 2 (2012) 284–298.
76. A. A. Zahir, A. Bagavan, C. Kamaraj, G. Elango and A. A. Rahuman, J. Biopest., 5 (2012) 95–102.
77. K. Shameli, M. B. Ahmad, S. D. Jazayeri, P. Shabanzadeh, P. Sangpour, H. Jahangirian, et al., Chem. Cent. J., 6 (2012) 73–95.
78. G. Geoprincy, B. N. V. Srri, U. Poonguzhali, N. N. Gandhi and S. Renganathan, Asian J. Pharm. Clin. Res., 6 (2013) 8–12.
79. C. A. Dos Santos, M. M. Seckler, A. P. Ingle, I. Gupta, S. Galdiero, M. Galdiero, A. Gade and M. Rai, J. Pharm. Sci., 103 (2014) 1931–1944.
80. S. S. Shankar, A. Ahmad and M. Sastry, Biotechnol. Prog., 19 (2003) 1627–1631.
81. S. S. Shankar, A. Rai, A. Ahmad and M. Sastry, J. Colloid Interf. Sci., 275 (2004) 496–502.
82. S. P. Chandran, M. Chaudhary, R. Pasricha, A. Ahmad and M. Sastry, Biotechnol. Prog., 22 (2006) 577–583.
83. J. Huang, Q. Li, D. Sun, Y. Lu, Y. Su, X. Yang, et al., Nanotechnology, 18 (2007) 105104–105114.
84. J. Y. Song and B. S. Kim, Korean J. Chem. Eng., 25 (2008) 808–811.
85. K. Jayendra, Y. K. Young, H. Jungho and R. Mahendra, J. Bionanosci., 3 (2009) 39–44.
86. M. Dubey, S. Bhadauria and B. S. Kushwah, Dig. J. Nanomater. Biostruct., 4 (2009) 537–543.
87. S. Vivekanandhan, M. Misra and A. K. Mohanty, J. Nanosci. Nanotechnol., 9 (2009) 6828–6833.

88. S. Vivekanandhan, D. Tang, M. Misra and A. K. Mohanty, Nanosci. Nanotechnol. Lett., 2 (2010) 240–243.

89. D. Philip, Physica E., 42 (2010) 1417–1424.

90. E. K. Elumalai, T. N. V. K. V. Prasad, V. Kambala, P. C. Nagajyothi and E. David, Arch. Appl. Sci. Res., 2 (2010) 76–81.

91. C. Krishnaraj, E. G. Jagan, S. Rajasekar, P. Selvakumar, P. T. Kalaichelvan and N. Mohan, Colloids Surf. B, 76 (2010) 50–56.

92. L. Lin, W. Wang, J. Huang, Q. Li, D. Sun, X. Yang, et al., Chem. Eng. J., 162 (2010) 852–858.

93. R. W. Raut, N. S. Kolekar, J. R. Lakkakula, V. D. Mendhulkar and S. B. Kashid, Nano-Micro Lett., 2 (2010) 106–113.

94. R. Sathyavathi, M. B. Krishna, S. V. Rao, R. Saritha and D. N. Rao, Adv. Sci. Lett., 3 (2010) 1–6.

95. A. K. Jha and K. Prasad, Int. J. Green Nanotechnol., 1 (2010) P110–P117.

96. V. Kumar, S. C. Yadav and S. K. Yadav, J. Chem. Technol. Biotechnol., 85 (2010) 1301–1309.

97. D. Philip, Spectrochim. Acta Part A, 78 (2011) 327–331.

98. C. Singh, V. Sharma, P. K. Naik, V. Khandelwal and H. Singh, Dig. J. Nanomater. Biostruct., 6 (2011) 535–542.

99. K. S. Mukunthan, E. K. Elumalai, T. N. Patel and V. R. Murty, Asian Pac. J. Trop. Biomed., 1 (2011) 270–274.

100. K. S. Prasad, D. Pathak, A. Patel, P. Dalwadi, R. Prasad, P. Patel, et al., Afr. J. Biotechnol., 10 (2011) 8122–8130.

101. M. Zargar, A. A. Hamid, F. A. Bakar, M. N. Shamsudin, K. Shameli, F. Jahanshiri, et al., Molecules, 16 (2011) 6667–6676.

102. K. Mallikarjuna, G. Narasimha, G. R. Dillip, B. Praveen, B. Shreedhar, C. S. Lakshmi, et al., Dig. J. Nanomater. Biostruct., 6 (2011) 181–186.

103. A. Kanchana, I. Agarwal, S. Sunkar, J. Nellore and K. Namasivayam, Dig. J. Nanomater. Biostruct., 6 (2011) 1741–1750.

104. L. Christensen, S. Vivekanandhan, M. Misra and A. K. Mohanty, Adv. Mater. Lett., 2 (2011) 163–167.

105. Y. S. G. Lakshmi, F. Banu, M. Ezhilarasan, D. Arumugam and D. Sahadevan, Green Synthesis of Silver Nanoparticles from Cleome viscosa: Synthesis and Antimicrobial Activity. Proceedings of International Conference on Bioscience, Biochemistry and Bioinformatics (IPCBEE), 5 (2011) IACSIT Press, Singapore.

106. C. Udayasoorian, K. Kumar and R. M. Jayabalakrishnan, Dig. J. Nanomater. Biostruct., 6 (2011) 279–283.

107. A. Parveen, A. S. Roy and S. Rao, Int. J. Appl. Biol. Pharm. Technol., 3 (2012) 222–228.

108. K. Mallikarjuna, G. R. Dillip, G. Narasimha, N. J. Sushma and B. D. P. Raju, Res. J. Nanosci. Nanotechnol., 2 (2012) 17–23.

109. S. Vivekanandhan, M. Venkateswarlu, D. Carnahan, M. Misra, A. K. Mohanty and N. Satyanarayana, Physica E., 44 (2012) 1725–1729.

110. D. Vishnudas, B. Mitra, S. B. Sant and A. Annamalai, Drug Invent. Today, 4 (2012) 340–344.

111. K. Rajathi and S. Sridhar, Int. J. Green Chem. Bioprocess, 2 (2012) 39–43.

112. M. Shivashankar and G. Sisodia, Int. J. Life. Sc. Bt. Pharm. Res., 3 (2012) 182–185.

113. S. Sulochana, P. Krishnamoorthy and K. Sivaranjani, J. Pharmacol. Toxicol., 7 (2012) 251–258.

114. M. J. Firdhouse and P. Lalitha, Asian J. Pharm. Clin. Res., 6 (2012) 92–94.

115. U. Mani, S. Dhanasingh, R. Arunachalam, E. Paul, P. Shanmugam, C. Rose and A. B. Mandal, Prog. Nanotechnol. Nanomater., 2 (2013) 21–25.

116. M. Lavanya, S. V. Veenavardhini, G. H. Gim, M. N. Kathiravan and S. W. Kim, Int. Res. J. Biol. Sci., 2 (2013) 28–34.
117. M. Vijayakumar, K. Priya, F. T. Nancy, A. Noorlidah and A. B. A. Ahmed, Ind. Crops Prod., 41 (2013) 235–240.
118. K. Gopinath, S. Gowri and A. Arumugam, J. Nanostruct. Chem., 3 (2013) 68–74.
119. S. Mukherjee, D. Chowdhury, R. Kotcherlakota, S. Patra, M. P. Bhadra, B. Sreedhar and C. R. Patra, Theranostics, 4 (2014) 316–335.
120. N. Gandhi, D. Sirisha and V. C. Sharma, J. Eng. Res. Appl., 4 (2014) 61–72.
121. M. Kumar, S. Dandapat, R. Ranjan, A. Kumar and M. P. Sinha, J. Microbiol. Exp., 6 (2018) 175–178.
122. S. Asha and P. Thirunavukkarasu, Acta Sci. Biol. Sci., 41,e45262 (2019) 1–9.
123. S. N. A. M. Yusuf, C. N. A. C. Mood, N. H. Ahmad, D. Sandai, C. K. Lee and V. Lim, R. Soc. Open Sci., 7,200065 (2020) 1–15.
124. S. Agarwal, M. Gogoi, S. Talukdar, P. Bora, T. K. Basumatary and N. N. Devi, RSC Adv., 10 (2020) 36686–36694.
125. V. V. Q. Bao, L. D. Vuong and Y. Wache, Nanomater. Energy, 10 (2021) 111–117.
126. J. L. Gardea-Torresdey, E. Gomez, J. R. Peralta-Videa, J. G. Parsons, H. Troiani and M. Jose-Yacaman, Langumir, 19 (2003) 1357–1361.
127. B. Ankamwar, C. Damle, A. Ahmad and M. Sastry, J. Nanosci. Nanotechnol., 5 (2005) 1665–1671.
128. D. Jain, H. K. Daima, S. Kachhwaha and S. L. Kothari, Dig. J. Nanomater. Biostruct., 4 (2009) 557–563.
129. N. Mude, A. Ingle, A. Gade and M. Rai, J. Plant Biochem. Biotechnol., 18 (2009) 83–86.
130. M. Sathishkumar, K. Sneha, S. W. Won, C. W. Cho, S. Kim and Y. S. Yun, Colloids Surf. B, 73 (2009) 332–338.
131. H. Bar, D. K. Bhui, G. P. Sahoo, P. Sarkar, S. P. De and A. Misra, Colloids Surf. A, 339 (2009) 134–139.
132. A. K. Singh, M. Talat, D. P. Singh and O. N. Srivastava, J. Nanopart. Res., 12 (2010) 1667–1675.
133. K. Vijayaraghavan, S. P. K. Nalini, N. U. Prakash and D. Madhankumar, Mater. Lett., 75 (2012) 33–35.
134. R. Geethalakshmi and D. V. L. Sarada, Int. J. Eng. Sci. Technol., 2 (2010) 970–975.
135. A. Saxena, R. M. Tripathi and R. P. Singh, Dig. J. Nanomater. Bios., 5 (2010) 427–432.
136. A. J. Jha and K. Prasad, Dig. J. Nanomater. Bios., 6 (2011) 1717–1723.
137. S. Kaviya, J. Santhanalakshmi, B. Viswanathan, J. Muthumary and K. Srinivasan, Spectrochim. Acta Part A, 79 (2011) 594–598.
138. A. P. Kulkarni, A. A. Srivastava, P. M. Harpale and R. S. Zunjarrao, J. Nat. Prod. Plant Resour., 1 (2011) 100–107.
139. S. Ankanna and N. Savithramma, Asn. J. Pharm. Clin. Res., 4, Suppl 2 (2011) 137–141.
140. S. A. Babu and H. G. Prabu, Mater. Lett., 65 (2011) 1675–1677.
141. G. D. G. Jobitha, C. Kannan and G. Annadurai, Drug Invent. Today, 4 (2012) 579–584.
142. G. VonWhite II, P. Kerscher, R. M. Brown, J. D. Morella, W. McAllister, D. Dean and C. L. Kitchens, J. Nanomater., Article ID 730746 (2012) 12 pages.
143. G. D. G. Jobitha, G. Annadurai and C. Kannan, Int. J. Pharm. Sci. Res., 3 (2012) 323–330.
144. M. Amin, F. Anwar, M. R. Janjua, M. A. Iqbal and U. Rashid, Int. J. Mol. Sci., 13 (2012) 9923–9941.

145. R. U. Maheswari, A. L. Prabha, V. Nandagopalan, V. Anburaja and J. J. Barath, J. Res. Nanobiotechnol., 1 (2012) 009–013.

146. M. Solgi and M. Taghizadeh, Int. J. Nanomater. Biostruct., 2 (2012) 60–64.

147. G. Gnanajobitha, M. Vanaja, K. Paulkumar, S. Rajeshkumar, C. Malarkodi, G. Annadurai and C. Kannan, Int. J. Nanomater. Biostruct., 3 (2013) 21–25.

148. T. Y. Suman, S. R. R. Rajasree, A. Kanchana and S. B. Elizabeth, Colloids Surf. B Biointerfaces, 106 (2013) 74–78.

149. S. S. Ravi, L. R. Christena, N. SaiSubramanian and S. P. Anthony, Analyst, 138 (2013) 4370–4377.

150. E. Rodríguez-León, R. Iñiguez-Palomares, R. E. Navarro, R. Herrera-Urbina, J. Tánori, C. Iñiguez-Palomares and A. Maldonado, Nanoscale Res. Lett., 8 (2013) 318–326.

151. G. Asghari, J. Varshosaz and N. Shahbazi, Nanomed. J., 1 (2014) 94–99.

152. S. Garg, A. Chandra, A. Mazumder and R. Mazumder, Asian J. Pharm., 8 (2014) 95–101.

153. Y. Gao, Q. Huang, Q. Su and R. Liu, Spectrosc. Lett., 47 (2014) 790–795.

154. K. R. Kavitha, S. Vijayalakshmi and B. M. Babu, Int. J. Recent Technol. Eng., 8 (2019) 5624–5629.

155. M. M. I. Masum, M. M. Siddiqa, K. A. Ali, Y. Zhang, Y. Abdallah, E. Ibrahim, W. Qiu, C. Yan and B. Li, Front. Microbiol., 10, 820 (2019) 1–18.

156. V. Parashar, R. Parashar, B. Sharma and A. C. Pandey, Dig. J. Nanomater. Biostruct., 4 (2009) 45–50.

157. N. Ahmad, S. Sharma, V. N. Singh, S. F. Shamsi, A. Fatma and B. R. Mehta, Biotechnol. Res. Inte., Article ID454090 (2011) 1–8.

158. C. Mason, S. Vivekanandhan, M. Misra and A. K. Mohanty, World J. Nano Sci. Eng., 2 (2012) 47–52.

159. S. C. G. K. Daniel, K. Nehru and M. Sivakumar, Curr. Nanosci., 8 (2012) 125–129.

160. S. Rajeshkumar, C. Kannan and G. Annadurai, Drug Invent. Today, 4 (2012) 511–513.

161. M. Intarasawang and K. Thamaphat, Adv. Mater. Res., 662 (2013) 80–83.

162. M. Khan, M. Khan, S. F. Adil, M. N. Tahir, W. Tremel, H. Z. Alkhathlan, A. Al-Warthan and M. R. H. Siddiqui, Int. J. Nanomed., 8 (2013) 1507–1516.

163. S. U. Ganaie, T. Abbasi, J. Anuradha and S. A. Abbasi, J. King Saud Univer., 26 (2014) 222–229.

164. Q. Chen, T. Mi, G. Chen and Y. Li, BioResources, 12 (2017) 7096–7106.

165. J. Compton, D. Thompson, D. Kranbuehl, S. Ohl, O. Gain, L. David and E. Espuche, Polymer, 47 (2006) 5303–5313.

166. J. Y. Lee, Y. Liao, R. Nagahata and S. Horiuchi, Polymer, 47 (2006) 7970–7979.

167. K. Mallick, M. J. Witcomb and M. S. Scurrell, Eur. Polym. J., 42 (2006) 670–675.

168. Z. Guo, H. Lin, A. B. Karki, S. Wei, D. P. Young, S. Park, J. Willis and T. H. Hahn, Compos. Sci. Technol., 68 (2008) 2551–2556.

169. X. Huang, P. Jiang, C. Kim, Q. Ke and G. Wang, Compos. Sci. Technol., 68 (2008) 2134–2140.

170. G. Zotti, B. Vercelli and A. Berlin, Chem. Mater., 20 (2008) 6509–6516.

171. S. Ghose, K. A. Watson, D. C. Working, J. W. Connell, J. G. Smith and Y. P. Sun, Compos. Sci. Technol., 68 (2008) 1843–1853.

172. A. Nyczyk, M. Hasik, W. Turek and A. Sniechota, Synth. Met., 159 (2009) 561–567.

173. T. I. Yang, R. N. C. Brown, L. C. Kempel and P. Kofinas, J. Nanopart. Res., 12 (2010) 2967–2978.

174. K. J. Jeon, H. R. Moon, A. M. Ruminski, B. Jiang, C. Kisielowski, R. Bardhan and J. J. Urban, Nat. Mater., March (2011) 1–5.

175. G. Karpacheva and S. Ozkan, Procedia Mater. Sci., 2 (2013) 52–59.
176. A. Heilmann, M. Quinten and J. Werner, Optical Response of Thin Plasma-Polymer Films with Non-Spherical Silver Nanoparticles, Eur. Phys. J. B, 3 (1998) 455–461.
177. Y. Zhou, S. Yu, C. Wang, Y. Zhu and Z. Chen, Chem. Lett., 28 (1999) 677–678.
178. A. B. R. Mayer and J. E. Mark, Polymer, 41 (2000) 1627–1631.
179. K. M. Raoa and M. Pattabi, J. New Mat. Electrochem. Systems, 4 (2001) 11–15.
180. I. Pastoriza-Santos and L. M. Liz-Marzán, Langmuir,18 (2002) 2888–2894.
181. D. Wu, X. Ge, Y. Huang, Z. Zhang and Q. Ye, Mater. Lett., 57 (2003) 3549–3553.
182. J. Y. Kim, D. H. Shin, K. J. Ihn and K. D. Suh, J. Ind. Eng. Chem., 9 (2003) 37–44.
183. Z. H. Mbhele, M. G. Salemane, C. G. C. E. van Sittert, J. M. Nedeljković, V. Djoković and A. S. Luyt, Chem. Mater, 15 (2003) 5019–5024.
184. K. Mallick, M. J. Witcomb and M. S. Scurell, J. Mater. Sci., 39 (2004) 4459–4463.
185. H. S. Shin, H. J. Yang, S. B. Kim and M. S. Lee, J. Colloid Interface Sci., 274 (2004) 89–94.
186. H. Huang and X. Yang, Carbohydr. Res., 339 (2004) 2627–2631.
187. H. Huang, Q. Yuan and X. Yang, Colloids Surf. B Biointerfaces, 39 (2004) 31–37.
188. J. H. Yeum, Q. Sun and Y. Deng, Macromol. Mater. Eng., 290 (2005) 78–84.
189. H. Takele, H. Greve, C. Pochstein, V. Zaporojtchenko and F. Faupel, Nanotechnology, 17 (2006) 3499–3505.
190. I. Prosyčevas, J. Puišo, A. Guobienė, S. Tamulevičius and R. Naujokaitis, Mater. Sci., 13 (2007) 188–192.
191. V. Thomas, M. M. Yallapu, B. Sreedhar and S. K. Bajpai, J. Colloid Interface Sci., 315 (2007) 389–395.
192. H. Kong and J. Jang, Biomacromolecules, 9 (2008) 2677–2681.
193. K. Vimala, K. S. Sivudu, Y. M. Mohan, B. Sreedhar and K. M. Raju, Carbohydr. Polym., 75 (2009) 463–471.
194. X. Huang, P. Jiang and L. Xie, Appl. Phys. Lett., 95 (2009) 242901-1–242901-3.
195. M. M. Kemp, A. Kumar, S. Mousa, T. J. Park, P. Ajayan, N. Kubotera, S. Mousa and R. J. Linhardt, Biomacromolecules, 10 (2009) 589–595.
196. Y. K. Mishra, S. Mohapatra, V. S. Chakravadhanula, N. P. Lalla, V. Zaporojtchenko, D. K. Avasthi and F. Faupel, J. Nanosci. Nanotechnol., 10 (2010) 2833–2837.
197. A. Tyurin, G. DeFilpo, D. Cupelli, F. P. Nicoletta, A. Mashin and G. Chidichimo, EXPRESS Polym. Lett., 4 (2010) 71–78.
198. S. W. Kang, Macromol. Res., 18 (2010) 705–708.
199. M. B. Ahmad, J. J. Lim, K. Shameli, N. A. Ibrahim and M. Y. Tay, Molecules, 16 (2011) 7237–7248.
200. M. B. Ahmad, M. Y. Tay, K. Shameli, M. Z. Hussein and J. J. Lim, Int. J. Mol. Sci., 12 (2011) 4872–4884.
201. A. M. Abdel-Mohsen, R. Hrdina and A. S. Aly, Adv. Chitin Sci., XIII (2011) 3–11.
202. K. Faghihi and M. Shabanian, J. Chil. Chem. Soc., 56 (2011) 665–667.
203. B. Sadeghi and A. Pourahmad, Adv. Powder Technol., (2011).
204. R. Shankar, U. Sahoo and V. Shahi, Macromolecules, 44 (2011) 3240–3249.
205. H. K. Chitte, N. V. Bhat, N. S. Karmakar, D. C. Kothari and G. N. Shinde, World J. Nano Sci. Eng., 2 (2012) 19–24.
206. Y. B. Wankhede, S. B. Kondawar, S. R. Thakare and P. S. More, Adv. Mat. Lett., 4 (2013) 89–93.
207. A. M. Abdelgawad, S. M. Hudson and O. J. Rojas, Carbohydr.Polym., 100 (2014) 166–178.
208. E. Hariprasad and T. P. Radhakrishnan, Langmuir, 29 (2013) 13050–13057.

209. G. F. Prozorova, A. S. Pozdnyakov, N. P. Kuznetsova, S. A. Korzhova, A. I. Emel'yanov, T. G. Ermakova, T. V. Fadeeva and L. M. Sosedova, Int. J. Nanomed., 9 (2014) 1883–1889.

210. G. Mustatea, I. Calinescu, A. Diacon and L. Balan, Mater.Plast., 51 (2014) 17–21.

211. Z. I. Ali, M. H. Helal, H. H. Saleh, A. F. Zikry and Y. A. Darwish, New York Sci. J., 7 (2014) 84–94.

212. C. M. T. Nguyen and V. T. Nguyen, Adv. Mater. Sci. Eng., Article ID 6650576 (2020) 1–9.

213. C. You, Q. Li, X. Wang, P. Wu, J. K. Ho, R. Jin, L. Zhang, H. Shao and C. Han, Sci. Rep., 7, 10489 (2017) 1–11.

214. A. L. Fadli, A. Hanifah, A. Fitriani, A. Rakhmawati and W. S. B. Dwandaru, International Conference on Science and Applied Science, AIP Conf. Proc. 2014, (2018) 020017-1–020017-7.

215. S. Pakseresht, A. W. M. Alogaili, H. Akbulut, D. Placha, E. Pazdziora, D. Klushina, Z. Konvičková, G. Kratošová, S. Holešová and G. S. Martynková, J. Nanosci. Nanotechnol., 19 (2019) 2938–2942.

216. M. A. Polinarski, A. L. B. Beal, F. E. B. Silva, J. Bernardi-Wenzel, G. R. M. Burin, G. I. B. de Muniz and H. J. Alves, Part. Part. Syst. Charact., 38 (2021) 2100009.

217. A. Regiel and A. Kyzioł, Chemik, 67 (2013) 683–692.

218. A. Regiel-Futyra, M. Kus-Liskiewicz, V. Sebastian, S. Irusta, M. Arruebo, A. Kyziol and G. Stochel, RSC Adv., 7 (2017) 52398–52413.

219. M. E. I. Badawy, T. M. R. Lotfy and S. M. S. Shawir, Bull. Natl. Res. Cent., 43 (2019) 1–14.

220. P. K. Dara, R. Mahadevan, P. A. Digita, S. Visnuvinayagam, L. R. G. Kumar, S. Mathew, C. N. Ravishankar and R. Anandan, SN Appl. Sci., 2, 665 (2020) 1–12.

221. H. Elfaig, W. Elsayed and H. Ahmed, MRS Adv., 5 (2020) 1331–1338.

222. S. A. Umoren, M. M. Solomon, A. Nzila and I. B. Obot, Materials, 13, 1629 (2020) 1–18.

223. J. W. Rhim, S. I. Hong, H. M. Park and P. K. W. Ng, J. Agric. Food Chem., 54 (2006) 5814–5822.

224. Y. K. Twu, Y. W. Chen and C. M. Shih, Powder Technol., 185 (2008) 251–257.

225. P. Sanpui, A. Murugadoss, P. V. Durga Prasad, S. Sankar Ghosh and A. Chattopadhyay, Int. J. Food Microbiol., 124 (2008) 142–146.

226. S. Lu, W. Gao and H. Y. Gu, Burns, 34 (2008) 623–628.

227. V. Thomas, M. M. Yallapu, B. Sreedhar and S. K. Bajpai, J. Biomater. Sci., Polym. Ed., 20 (2009) 2129–2144.

228. D. K. Božanić, L. V. Trandafilović, A. S. Luyt and V. Djoković, React. Funct. Polym., 70 (2010) 869–873.

229. D. vanPhu, V. T. K. Lang, N. T. K. Lan, N. N. Duy, N. D. Chau, B. D. Du, B. D. Cam and N. Q. Hien, J. Exp. Nanosci., 5 (2010) 169–179.

230. S. Honary, K. Ghajar, P. Khazaeli and P. Shalchian, Trop. J. Pharm. Res., 10 (2011) 69–74.

231. J. An, Q. Luo, X. Yuan, D. Wang and X. Li, J. Appl. Polym. Sci., 120 (2011) 3180–3189.

232. S. Govindan, E. A. K. Nivethaa, R. Saravanan, V. Narayanan and A. Stephen, Appl. Nanosci., 2 (2012) 299–303.

233. P. Kaur, A. Choudhary and R. Thakur, Int. J. Sci. Eng. Res., 4 (2013) 869–872.

234. A. M. Youssef, M. S. Abdel-Aziz and S. M. El-Sayed, Int. J. Biol. Macromol., 69 (2014) 185–191.

235. M. A. Alghuthaymi, K. A. Abd-Elsalam, A. Shami, E. Said-Galive, E. V. Shtykova and A. V. Naumkin, J. Fungi, 6, 51 (2020) 1–18.

236. M. S. Al-saggaf, Int. J. Polym. Sci., Article ID 5578032 (2021) 1–9.

237. S. Raza, A. Ansari, N. N. Siddiqui, F. Ibrahim, M. I. Abro and A. Aman, Sci. Rep., 11, 10500 (2021) 1–15.

238. P. Dwivedi, S. S. Narvi and R. P. Tewari, Int. J. Green Nanotechnol., 4 (2012) 248–261.

239. P. Dwivedi, S. S. Narvi and R. P. Tewari, Adv. Mater. Res., 585 (2012) 144–148.

240. P. Dwivedi, S. S. Narvi and R. P. Tewari, Ind. Crops Prod., 54 (2014) 22–31.

241. P. Dwivedi, S. S. Narvi and R. P. Tewari, International Conference on Nanoscience, Technology and Societal Implications, NSTSI11 (2011).

242. P. Dwivedi, S. S. Narvi and R. P. Tewari, Int. J. Biomed. Nanosci. Nanotechnol., 2 (2012) 187–206.

243. P. Dwivedi, S. S. Narvi and R. P. Tewari, J. Chin. Med. Res. Dev., 1 (2012) 23–27.

244. P. Dwivedi, S. S. Narvi and R. P. Tewari, Int. J. Adv. Eng., Sci. Technol., 2 (2012) 236–243.

245. P. Dwivedi, S. S. Narvi and R. P. Tewari, Int. J. Eng. Res. Appl., 2 (2012) 1490–1495.

246. P. Dwivedi, S. S. Narvi and R. P. Tewari, Int. J. Sci. Res. Publ., 2 (2012) 1–5.

247. P. Dwivedi, S. S. Narvi and R. P. Tewari, Ann. Res. Rev. Biol., 4 (2014) 1059–1069.

248. P. Dwivedi, S. S. Narvi and R. P. Tewari, Adv. Sci. Eng. Med., 6 (2014) 1–9.

249. P. Dwivedi, S. S. Narvi and R. P. Tewari, Nano LIFE, 5 (2015) 1540006.

250. P. Dwivedi, D. Tiwary, S. S. Narvi, R. P. Tewari and K. P. Shukla, Lett. Appl. Nanobiosci., 9 (2020) 1485–1493.

251. P. T. Anastas and J. C. Warner, Green Chem., Oxford University Press: New York, NY, (1998).

3 Phytochemicals-Aided Conversion to Silver Nanoparticles
Nanobiotechnology

3.1 INTRODUCTION

Hitherto, nanoparticles of a range of metals and metal oxides have boomed [1–3], but silver nanoparticles have bloomed [4–6] and captured special attention in the era with a wide variety of specific applications, especially in the biomedical arena, encompassing broad spectrum of distinct purposes [7–14]. Silver since centuries has been recognized for being possessed with strong inherent antimicrobial characteristics having inhibitory effect even against multidrug resistant human pathogens [15]. Silver containing antimicrobial materials capture much attention because of the non-toxicity of the active silver ions (Ag^+) to human cells as well as for their novelty of being a much lasting biocide with low volatility and high temperature stability [16]. The antimicrobial and antiseptic properties of Ag^+ have been significant, with only few bacteria may be still intrinsically resistant. Thus, silver is a potential resource for therapy in medicine and uses in biomedical engineering [17, 18]. The antimicrobial activity of silver relies on the Ag^+, binding strongly to electron donor groups available in biological molecules. Silver nanoparticles having high surface to volume ratio act more effectively and exhibit superior properties than the bulk metal, therefore are in great demand in medicine and biomedical industry.

So, in this recently innovated active area of research, by blending herbalism with nanotechnology, we present an effort to analyze the similarity together with the variability in the morphological, physical and chemical nature of the silver nanostructures, which are mediated and controlled by different herbal plant materials with varying property in their metabolites. We have hand-picked few plants from our botanical diversity [19] and proceeded for the primarily qualitative assessment as well as comparative analysis of the fabricated silver nanoparticles through the nanobiotechnology route, aided by the extracts of specific plant materials (phytomass) mentioned here. They are dried fruits or endocarp with blueberry seeds of the plant *Elaeocarpus ganitrus* Roxb. (Rudraksha beads); bark, foliage and fruits of *Terminalia arjuna* Roxb. (Arjuna); foliage of *Pseudotsuga menziesii* (Christmas tree); *Prosopis spicigera* (Shami); *Ficus religiosa* (Peepal or sacred Fig); *Ocimum sanctum* (Tulsi) and rhizome of *Curcuma longa* (Turmeric).

DOI: 10.1201/9781003217343-3

3.1.1 ETHNOBOTANY

Elaeocarpus ganitrus — **Rudraksha**

- Widely known according to Ayurvedic and Unani medical systems, Rudraksha possesses immense medicinal value together with vital biomedical, electromagnetic and inductive properties, therefore regarded as the king of herbal medicines [20–24].
- The five-faceted (panchmukhi) Rudraksha bead is believed to have tremendous healing power and revered to as a remedy, for the treatment and benefactor, of many mental diseases and physical ailments.

Terminalia arjuna — **Arjun**

- It is cardiac stimulant; strengthens heart muscles, heart functionality, having prostaglandin inducing and coronary risk modulating properties.
- An astringent and hemostatic; review reveals that *T. arjuna* is a very influential plant for its extensive phytochemicals and pharmacological properties consisting medicinally important chemicals [25].

Pseudotsuga menziesii — **Christmas tree**

- Important as Christmas decorations; decorations when removed can have fruitful utilization.
- Needles are loaded with vitamins A and C. The vitamin C levels are very high (approximately five times greater than lemons). They also have flavonoids, quercetin, anthocyanins, proanthocyanidins, resveratrol, tannins and much more.
- Vitamin C and proanthocyanidins are considered to be important in preventing cancer.
- Vitamin C is a coenzyme in many enzymatic reactions, significantly important in wound healing and in preventing capillary bleeding [26].
- The properties of flavonoids mainly are antioxidant, antihistaminic, anti-inflammatory and antiviral.
- Tannins act as astringent.

Prosopis spicigera — **Shami**

- All parts of the tree are useful, so it is called kalp taru and also called the 'king of desert', or 'wonder tree'.
- Water-soluble extract of the stem bark shows anti-inflammatory properties [27].

Ficus religiosa — **Peepal**

- In traditional medicine *Ficus religiosa* is for the cure of more than 50 types of disorders including epilepsy, asthma, diabetes, diarrhea, gastric, inflammatory, infectious and sexual disorders [28].

Ocimum sanctum — Tulsi/Basil

- The holy basil has been considered among the India's most powerful herbs, since thousands of years. The consistent use of this herb has far-reaching effects in maintaining the balance of the energy centers (*chakras*) of the human body. It is designated as a 'rasayana' herb which nourishes the person's growth to perfect health and promotes long life [29].

Curcuma longa — Turmeric

- A small perennial herb native to India, bearing rhizomes on its root system which are used as a culinary spice known as Turmeric, and Curcumin is its chief medicinal component.
- Curcumin is able to break up the amyloid-beta polymers in Alzheimer's, suppress arthritic inflammation, induce apoptosis in cancer cells and enhance insulin sensitivity.
- Apart from it, *Curcuma longa* also has a wide range of medicinal benefit in particular with regard to its secondary metabolites as sources of hepatoprotective, antioxidants, anti-inflammatory, anticarcinogenic, antimicrobial, for cardiovascular effect, gastrointestinal effect and enhances immunity due to curcumin – the active constituent of turmeric [30].

3.1.2 PHYTOCHEMISTRY OF THE AQUEOUS EXTRACTS

Elaeocarpus ganitrus — Rudraksha: bead (dried fruit – endocarp with seeds enclosed)

- Phytochemical screening indicates the availability of phytosterols, polyphenolic compounds, tannins (gallic and ellagic acids), fatty acids (palmitic and linoleic acids), alkaloids, flavonoids (proanthocyanidins and quercetin), carbohydrates, glycosides and proteins in the aqueous extract of the Rudraksha bead [31–34].

Terminalia arjuna — Arjun: bark, foliage, fruit

- Bark: Chemical analysis of the water extract of the bark, confirmed presence of sugar, coloring agents, tannins, glycosides, saponins, flavonoids together with inorganic salts like carbonates of sodium, calcium and chlorides of alkali metals in traces. The main active constituents are triterpenoids saponins (terminic acid, arjunolic acid, arjunic acid, arjunin and arjungenin), glycosides (arjunetin, arjunoside, arjunaphthalonoside and terminoside), having cardioprotective property [25], while flavonoids (arjunolone, arjunone, quercetin and proanthocyanidins) act as antihistamine having anti-inflammatory and anti-allergic actions, together with antioxidant property, whereas tannins (pyrocatechols, terflavin C and gallic acid) have astringent, wound healing and antimicrobial activities [35–37].

- Foliage: Chief chemical constituents of the aqueous extract are tannins, alkaloids, steroids, carbohydrates, sparing amounts of saponins and flavonoids [38].
- Fruit: Aqueous fruit extract contains tannins, terpenoids, saponins, flavonoids, glycosides and polyphenolic compounds [39].

Pseudotsuga menziesii — Christmas tree: foliage

- Needles are loaded with high levels of vitamin C, which doesn't end there, will not be destroyed if extracted into boiling water. The solubility of vitamin C in water is 80% at 100°C, while there is only some decomposition at 190–192°C which is its melting point and this is a temperature that is not reached from boiling water, therefore much of that vitamin C content can be taken into benefit of improving health to a certain degree.
- Vitamin C, a glucose-related six carbon compound, the biologically active form of ascorbic acid, considered an antioxidant, has the function of a good reducing agent while acts as a coenzyme in many metabolic pathways [40].
- Aqueous extract also contains carbohydrates, tannins with astringency, anti-allergic flavonoids (proanthocyanidins, anthocyanins and sparing amounts of quercetin), etc., which may offer a number of benefits to humans [41, 42].

Prosopis spicigera — Shami: foliage

- There is presence of alkaloids, carbohydrates, proteins/amino acids, glycosides, tannins, terpenoids, flavonoides, saponins and sterols.
- A large number of amino acids isolated from the leaves are glutamic acid, aspartic acid, glycine, serine, arginine, alanine, histidine, proline, tyrosine, threonine, valine, methionine, cysteine, leucine, isoleucine, lysine and phenylalanine [43].

Ficus religiosa — Peepal: foliage

- The phytochemical evaluation of the aqueous leaf extract gave the presence of alkaloids, flavonoids, saponins, phenols, tannins, terpenoids, cardiac glycosides, phytosterols, resins carbohydrates and amino acids, such as arginine, serine, aspartic acid, glycine, threonine, alanine, proline, tryptophan, tryosine, methionine, valine, isoleucine, and leucine [44].

Ocimum sanctum — Tulsi/Basil: foliage

- Aqueous leaf extract contains a variety of constituents and nutrients having highly complex chemical composition and biological activity.
- Carbohydrates, saponins, terpenenoids (eugenoland linalool), phenolic compounds, such as tannins, flavonoids (orientin, vicenin, luteolin, rosmarinic acid and anthocyanins), are abundant in Tulsi plant extract.

The two water-soluble flavonoids (orientin and vicenin) have got strong antioxidant and antcarcinogenic activity. While the compound eugenol, slightly soluble in water, has numerous huge health benefits [45].

Curcuma longa — Turmeric: rhizome

- The primary nutrients and secondary metabolites screening shows *Curcuma longa* rhizome aqueous extract is rich in turmerin (a water-soluble antioxidant/DNA-protectant/antimutagenpeptide) [46], amino acids, fatty acids, flavonoids, alkaloids, triterpenoids, glycosides, reducing sugars and leaking of anthocyanins, anthraquinones, coumarins, emodins, lignins, leuco anthocyanins, phenols, saponins, steroidsand tannins [47, 48] with sparing amounts of curcumin [49].

3.1.3 MECHANISM OF PHYTOMASS-MEDIATED CONVERSION

The mechanism of silver nanoparticle synthesis is bio-reduction of silver ions (Ag^+) to silver metal atoms (Ag^0) followed by colloidal aggregation and self-assembling [50, 51]. Organic phytochemicals are the chief biomolecules principally responsible for mediating the reaction of nanoparticle generation, by simultaneously performing the dual action of reducing agents and capping ligands. The plant metabolites furthermore function as post-synthesis stabilizers too for the phytofabricated nanosilver thereafter. On the whole the biosynthesis of silver nanoparticles is conducted at ambient temperature and the entire process proceeds in a facile manner.

Through Fourier transformed infrared (FTIR) spectroscopy, it has been indicated that plant metabolites, mainly sugars, as well as amino acids, enzymes, vitamins, terpenoids, alkaloids, polyphenols, phenolic acids and other phenolic compounds, play vital role in reducing Ag^+ to Ag^0 and subsequently supporting their stability into silver nanoparticles [52–55]. Based on the FTIR spectroscopy data, it has been postulated [56] that a proton dissociation of the OH-group from phenolic compounds forms resonance structures which are capable of further oxidation. This is followed by the active reduction process of Ag^+ and formation of nanoparticles. It has also been assumed that tautomeric transformations of flavonoids (a large group of polyphenolic compounds) from the enol to keto form release a reactive hydrogen atom capable of reducing metal ions and play a key role to form nanoparticles [57]. A few flavonoids with their π-electrons or carbonyl groups can even chelate metal ions. Example quercetin a flavonoid with very good chelating character can chelate by several positions through the carbonyl and hydroxyls present at the C3 and C5 positions whereas with the catechols at C3' and C4' sites. Interestingly, such mechanisms can only justify flavonoids being adsorbed on the surface of phytofabricated nascent nanoparticles.

FTIR data evaluation of nanoparticles fabricated through plant extracts elucidates that nascent nanoparticles are also very commonly found in association with functional groups of amino acids [58]. Currently, the 20 natural α-amino acids have been analyzed to identify their potential for reducing and binding to metal ions. It has been observed that different amino acids have differential capability to reduce metal ions and bind to them [59]. Amino acids, such as arginine, cysteine, methionine and lysine, can generally bind to Ag^+ [60], even though it is not specific and others can also bind. Amino acids usually bind to metal ions through the carboxylate and amino groups of their main or side chains, e.g. the carboxyl groups of glutamic and aspartic acid, or even through a nitrogen atom of the histidine imidazole ring. Other examples through which side chains bind to metal ions are thioether (methionine), thiol (cysteine), carbonyl groups (asparagine and glutamine) and hydroxyl (serine, threonine and tyrosine) [61]. Studies have showed that the carbonyl groups of glutamine and asparagines, hydroxyl groups of tyrosine and thiol groups of cysteine along with other amino groups can also be involved in the Ag^+ reduction process.

According to several FTIR data, the sugars/carbohydrates present in plant extracts can be considered as the chief inducers for the fabrication of metal nanoparticles. A well-known fact is that monosaccharides, like glucose (linear with an aldehyde group), are good reducing agents [62]. Glucose is usually in its cyclic form but being in equilibrium with the linear form containing an aldehyde group. Disaccharides and polysaccharides also possess reducing ability, which relies on the ability of their individual monosaccharide components to achieve an open chain form in an oligomer so that a metal ion may access an aldehyde group. Therefore, disaccharides, such as lactose and maltose, are also reducing agents because one of their monomers can at least be able to assume open chain. Postulation is also made that, via nucleophilic addition of OH^- nucleophile, the sugar aldehyde group undergoes oxidation into carboxyl group and this results in reduction of the metal ions that leads to the formation of metal nanoparticles [52].

Sugars also have many C-OH functional groups of alcohol, which are oxidized and converted into the C=O group of an aldehyde.

Oxidation of alcohol groups:

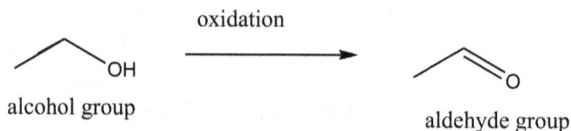

alcohol group   aldehyde group

Oxidation half reaction:

$$-CH_2CH_2OH \rightarrow -CH_2CHO + 2e^- + 2H^+$$

Reduction half reaction:

$$2Ag^+ + 2e^- \rightarrow 2Ag^0$$

The -CHO group of aldehydes are much easier to be oxidized into the -COOH carboxylic groups.

Oxidation of aldehyde groups:

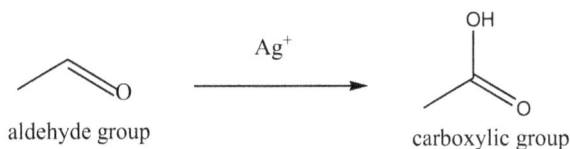

aldehyde group   carboxylic group

Oxidation half reaction:

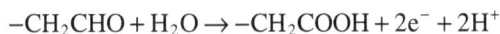

$$-CH_2CHO + H_2O \rightarrow -CH_2COOH + 2e^- + 2H^+$$

Reduction half reaction:

$$2Ag^+ + 2e^- \rightarrow 2Ag^0$$

Net reaction:

$$-CH_2CHO + H_2O + 2Ag^+ \rightarrow -CH_2COOH + 2H^+ + 2Ag^0$$

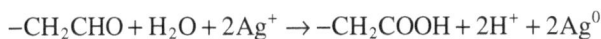

Any compound or sugar like glucose with an aldehyde group will react with Ag⁺ to give Ag⁰.
  Partial oxidation reactions of reducing sugars (glucose):

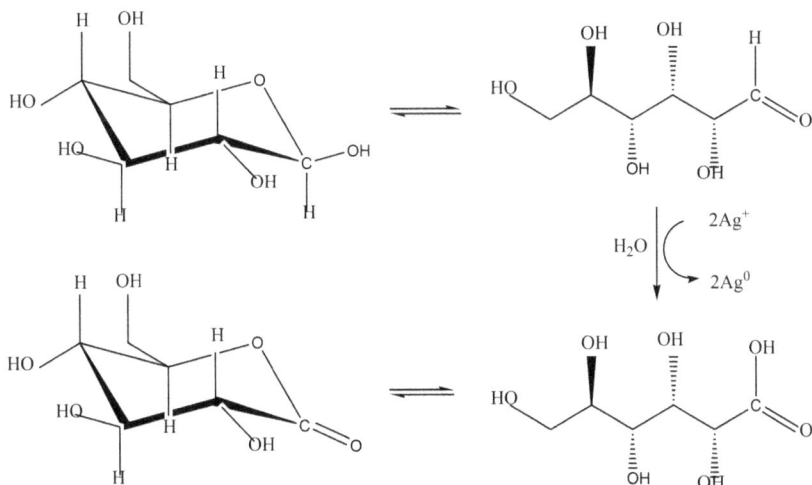

Furthermore, it has also been presumed that biosynthetic by-products or the reduced coenzymes and cofactors have a crucial role in the reduction of metal salts to the respective nanoparticles. Therefore, as has been investigated, the effect of primarily glucose (as the electron donor) is highly responsible in the process of nanoparticle synthesis and bio-reduction of silver ions [63]. Hypothetically, electron donors such as glucose may first provide electrons for the reduction of coenzymes such as nicotinamide adenine dinucleotide (NAD$^+$ $\leftrightarrow$ NADH), together with its close analog nicotinamide adenine dinucleotide phosphate (NADP$^+$ $\leftrightarrow$ NADPH), cofactors like flavin adenine dinucleotide (FAD $\leftrightarrow$ FADH2), and then make the reduction of silver ions (Ag$^+$) to silver metal atoms (Ag^0). Then probably this reduction of Ag$^+$ can be possible through the mechanism of glycolysis [64]. Glycolysis is the known biological route that basically converts glucose (C$_6$H$_{12}$O$_6$) into pyruvate (CH$_3$COCO$_2^-$) and hydrogen ion (H$^+$) through a definite series of ten reactions involving a set of ten intermediates. The free energy that is released in this sequential process is harvested, forming the high-energy compound called adenosine triphosphate (ATP), while at the same time is produced the reduced form of nicotinamide adenine dinucleotide (NADH), a strong reducing agent.

Representing the following equations illustrates in brief:

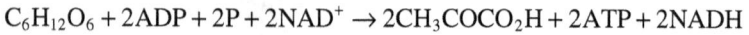

$$C_6H_{12}O_6 + 2ADP + 2P + 2NAD^+ \rightarrow 2CH_3COCO_2H + 2ATP + 2NADH$$

$$2NADH + 2H^+ \rightarrow 2NAD^+ + 4e^- + 4H^+$$

$$4AgNO_3 \rightarrow 4Ag^+ + 4NO_3^-$$

$$4Ag^+ + 4e^- \rightarrow 4Ag^0$$

The H$^+$ and NO$_3^-$ ions are released to the aqueous solvent.

Large amounts of protons and electrons are also produced in simple hydrolysis of glucose which can reduce Ag$^+$.

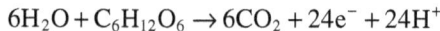

$$6H_2O + C_6H_{12}O_6 \rightarrow 6CO_2 + 24e^- + 24H^+$$

$$24AgNO_3 \rightarrow 24Ag^+ + 24NO_3^-$$

$$24Ag^+ + 24e^- \rightarrow 24Ag^0$$

However, the overall phytofabrication mechanism of silver nanoparticle synthesis by plant extracts may be divided mainly into three phases: (1) the phase of activation, when there is reduction of Ag$^+$ to Ag^0 and nucleation of the metal atoms; (2) the phase of growth, when the adjacent small nanoparticles spontaneously merge resulting into larger sized particles or also by heterogeneous nucleation manner there is direct increase in nanoparticle size and growth with further reduction of metal ions, i.e. the process of Ostwald ripening, which increases the thermodynamic stability of the formed nanoparticles; and (3) the phase of termination, which finally determines the shape and size of the silver nanoparticles [61, 65]. It is in this final phase, when energetically most favorable conformation is acquired by the nanoparticles. This process is intensely influenced by the plant phytochemicals and their capability of stabilizing the nanoparticles. Therefore, nanotriangles having a very high-surface energy are less stable and need to be supported by the biomolecules present, or else they may acquire a more stable morphology, e.g. of a truncated triangle, in order to have minimum 'Gibbs free surface energy'.

Hence, control over the morphology, like shape and size of the nanoparticles, can also be linked, to the interaction of the plant biomolecules present, with the metal ions [53]. The lone pair of electrons present on the nitrogen moiety or other such groups acting as Lewis base to the partial positive charge created and present on the surface of the silver nanoparticles, due to slight electron drift, does the stabilization. Different plants and their parts may differ in the composition and concentration of these active biological components. This reasonably explains, to some extent, the obtained diversity in morphology of the phytofabricated silver nanoparticles: e.g. spheres, ellipsoids, triangles, cubes, pentagons, hexagons, nanorods and nanowires [66].

3.2 EXPERIMENTAL SECTION

3.2.1 MATERIALS

Plant materials: foliage of *Pseudotsuga menziesii, Prosopis spicigera, Ficus religiosa, Ocimum sanctum*; bark, foliage and fruits of *Terminalia arjuna* Roxb. and rhizome of *Curcuma longa* were collected freshly from the surroundings. Five-faced dried fruits (endocarp with seeds enclosed) of *Elaeocarpus ganitrus* Roxb., i.e. the beads of Rudraksha, were purchased from Rudraksha World, Prayagraj, India. All the samples of plant materials were identified and authenticated by Dr. Akhauri Pr. Sahay, Dept. of Botany, B. R. A. Bihar University.

Analytical grade silver nitrate ($AgNO_3$), nutrient broth and nutrient agar were procured from Thomas Baker (Chemical) Pvt. Ltd., India. Bacterial strains *Pseudomonas aeruginosa* (Gram negative) and *Staphylococcus aureus* (Gram positive) were obtained from the culture bank of Department of Microbiology, Sam Higginbottom Institute of Agriculture, Technology and Sciences, Allahabad, India. All solutions required were prepared using deionized water.

3.2.2 DIVERSE PLANT EXTRACT PREPARATION: FROM *ELAEOCARPUS GANITRUS* ROXB., *TERMINALIA ARJUNA* ROXB., *PSEUDOTSUGA MENZIESII*, *PROSOPIS SPICIGERA*, *FICUS RELIGIOSA*, *OCIMUM SANCTUM* AND *CURCUMA LONGA*

The plant materials, mentioned and depicted in Figures 3.1–3.9, were washed, air dried and weighed to obtain the phytomass. All the plant materials were cut into fine pieces, except the Rudraksha beads to be used as whole. Phytomass extracts of different plant materials were prepared separately by boiling in 100 mL of sterile deionized water in 500 mL Erlenmeyer flask at 100°C for 15 min and 30 min for the Rudraksha beads. The crude plant extracts were filtered using filter paper of Whatman No. 41 and stored in closed bottles thereafter at 4°C for further requirement.

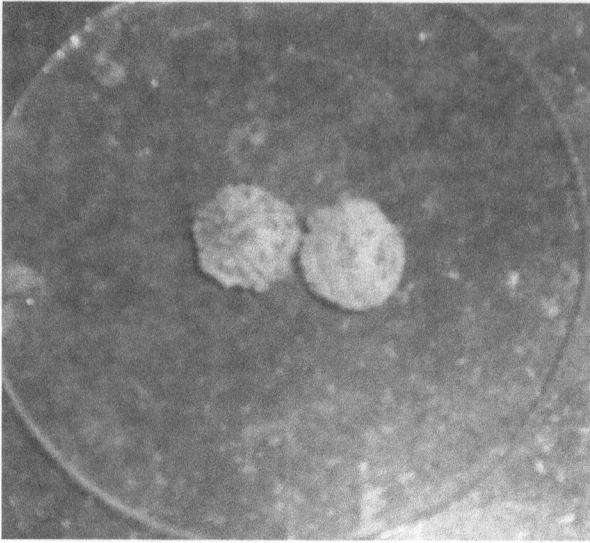

FIGURE 3.1   Seeds/beads of *Elaeocarpus ganitrus*.

FIGURE 3.2   Bark of *Terminalia arjuna*.

FIGURE 3.3 Foliage of *Terminalia arjuna*.

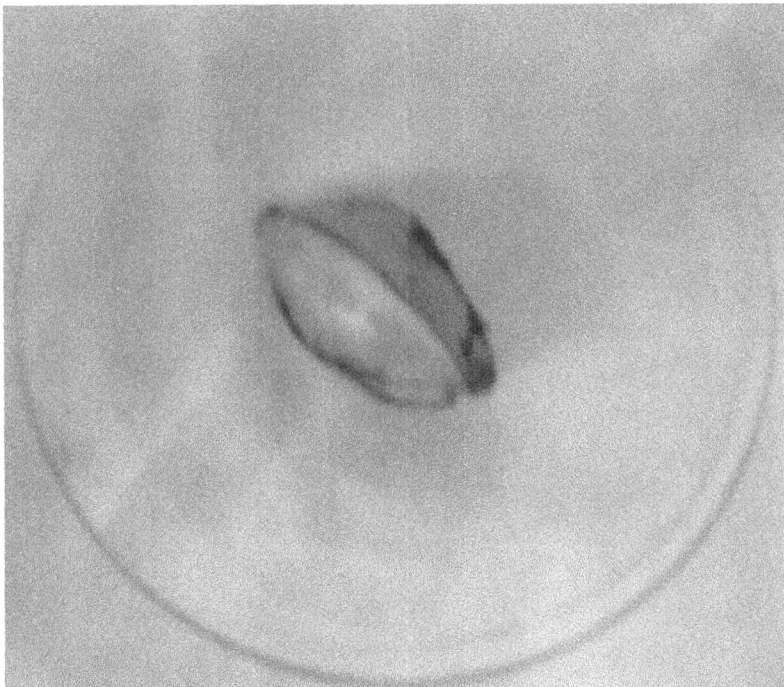

FIGURE 3.4 Fruit of *Terminalia arjuna*.

FIGURE 3.5　Foliage needles of *Pseudotsuga menziesii*.

FIGURE 3.6　Foliage of *Prosopis spicigera*.

FIGURE 3.7 Foliage of *Ficus religiosa.*

FIGURE 3.8 Foliage of *Ocimum sanctum.*

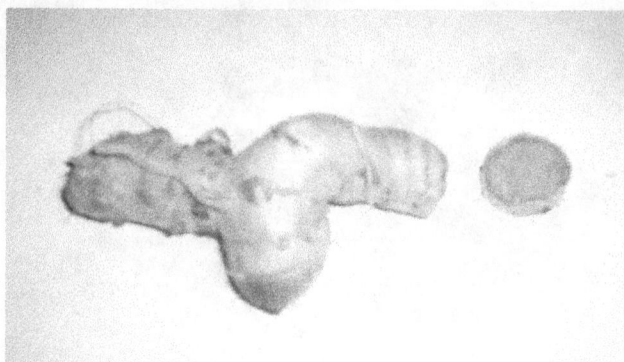

FIGURE 3.9   Rhizome of *Curcuma longa*.

Phytomass extracts consisting plant biomolecules were labeled as follows: PE1, PE2, PE3, PE4, PE5, PE6, PE7, PE8 and PE9 for the seeds of *Elaeocarpus ganitrus* (Rudraksha beads), bark of *Terminalia arjuna*, foliage of *Terminalia arjuna*, fruits of *Terminalia arjuna*, foliage needles of *Pseudotsuga menziesii*, foliage of *Prosopis spicigera*, foliage of *Ficus religiosa*, foliage of *Ocimum sanctum* and rhizome of *Curcuma longa*, respectively.

3.2.3 PHYTOMASS CONVERSION OF SILVER TO SILVER NANOPARTICLES

Rudraksha extracts 50 mL, while Turmeric extracts 10 mL, and then 5 mL each of the aqueous phytomass extracts of other plant materials were added to 100 mL of 1 mM silver nitrate ($AgNO_3$) solutions separately. All the reaction mixtures were kept in closed bottles, labeled clearly as RM1, RM2, RM3, RM4, RM5, RM6, RM7, RM8 and RM9 according to the phytomass extracts PE1, PE2, PE3, PE4, PE5, PE6, PE7, PE8 and PE9 added to them, respectively. The reaction mixtures were allowed to incubate at room temperature (RT) for complete stabilization and further observation. The entire initial experimental work carried out in the laboratory for the phytofabrication of silver nanoparticles is specified and summarized in Table 3.1.

3.2.4 UV-VISIBLE ABSORBANCE SPECTROSCOPY

The reduction of Ag^+ to Ag^0 was monitored through spectral data obtained from UV-Vis. spectrophotometer (Shimadzu UV 2450), which were recorded ranging between 200 and 800 nm, of aliquots from each reaction mixture (containing $AgNO_3$ solution+ plant extract) after 48 h when complete stabilization had occurred and no further color transformation was observed. Spectra of the separate plant extracts alone and also of $AgNO_3$ solution were recorded. The aliquots of the test samples were diluted '4×' according to requirement. Deionized water was used as blank.

TABLE 3.1

Qualitative and Quantitative Result of Selected Plant Materials (Phytomass) in the Color Transformation Process of the Different Labeled Reaction Mixtures after the Addition of Phytomass Extracts to Transparent and Colorless AgNO₃ Solution

S N RM	Phytomass for Reaction Mediation	Weight of Phytomass (g)	Color of Phytomass Extract	Volume of Phytomass Extract Added (mL)	Time duration for Initiation of Color Transformation after Addition of Phytomass Extract	Time Duration Demanding Complete Color Transformation Denoting Stabilization of Reaction Mixture	Rate of Color Transformation	Final Color Developed and Color Intensity
1	*Elaeocarpus ganitrus* (Rudraksha) beads	3.91	Transparent without any color	50	~1 h	~24 h	Gradual	Olive brown–khaki
2	*Terminalia arjuna* (Arjuna) bark	20.99	Crimson red	5	~3 min	~30 min	Rapid	Deep brown
3	*Terminalia arjuna* (Arjuna) foliage	20.98	Lemon green	5	Instant	~15 min	Particularly rapid	Intense dark brown
4	*Terminalia arjuna* (Arjuna) fruit	21.01	Orangish green	5	~5 min	~45 min	Rapid	Darkish brown
5	*Pseudotsuga menziesii* (Christmas tree) foliage	19.11	Vintage pale green	5	~30 min	~2 h	Moderately rapid	Orangish brown
6	*Prosopis spicigera* (Shami) foliage	21.12	Lemon green	5	Instant	~15 min	Distinctly rapid	Intense dark brown
7	*Ficus religiosa* (Peepal) foliage	21.11	Light green	5	~10 min	~2 h	Moderately rapid	Deep brown
8	*Ocimum sanctum* (Tulsi) foliage	19.13	Original mauve	5	Instant	~15 min	Extremely rapid	Intense dark brown
9	*Curcuma longa* (Turmeric) rhizome	21.21	Bright turmeric yellow	10	~24 h	~21 days	Steady	Orangish brown

3.2.5   TEM Observations and Nanoparticle Size Analysis

Transmission electron microscopy (TEM) samples of the biosynthesized silver nanoparticles were prepared by placing drops of the suspensions onto carbon-coated copper grids and allowing the solvent to evaporate. Observations were performed with the help of (HR TEM TECNAI 20 G^2) instrument operated at 200 kV accelerating voltage. Particle size analysis was done with (NANOTECH) particle size analyzer instrument.

3.2.6   Chemical Composition Identification

The chemical composition identification of the nanoparticles was done by energy dispersive X-ray spectroscopic analysis (EDAX), performed with (HR TEM TECNAI 20 G^2) instrument operated at an accelerating voltage of 200 kV.

3.2.7   Antimicrobial Assay

After complete stabilization, all the reaction mixtures were also centrifuged separately at 9,000 rpm for 15 min and the residues of silver nanoparticles obtained were re-dispersed in distilled water. This procedure was repeated three times to isolate silver nanoparticles from loosely bound bio-organic molecules or other compounds present. The remnant residues obtained as such were assayed for antimicrobial activity against *S. aureus* (Gram positive) and *P. aeruginosa* (Gram negative). Disc diffusion method was the preferred protocol finding out the standard zone of inhibition (ZOI). Wells were bored into disc shape having ~7 mm diameter on different cultured agar plates, and placed the nanosilver residues. For positive and negative control, 400 mg of ≥98.0% Sparfloxacin powder (C+) having 11 mm disc diameter and distilled water (C−) were used, respectively. The culture media used was nutrient agar and inoculation was with 1 mL of the bacterial organism containing broth. These plates containing the bacterial and silver nanoparticles were incubated at 37°C for 48 h. The plates were then examined for evidence of ZOI, which appear as a clear area around the disc. The diameters of such ZOI were measured using a meter ruler.

3.2.8   In vitro Cytotoxicity Testing

In vitro cytotoxicity test was done through MTT assay against the J774A.1 murine macrophage cell line. Silver nanoparticle suspensions from RM5, RM7 and RM9 were selectively taken and placed as the test samples for determination and comparison of the toxicity effects between samples having high and low antimicrobial activity. Cells were placed in a 96-well micro plate at a density of 2.5×10^4 cells/well. Then the cells were incubated in triplicate with varying concentrations for 72 h at 37°C, 5% CO_2 and untreated cells were served as control. Cytotoxicity assessment was through the colorimetric MTT transformation assay. The 50% cytotoxicity concentration (CC_{50}) was then calculated, while the graph plotted showing the optical density (OD) against drug concentration, OD of the control well taken as 100% survival.

3.3 RESULTS AND DISCUSSION

3.3.1 EFFECT OF VARIATION IN PHYTOCHEMICALS ON SILVER NANOPARTICLE SYNTHESIS OBSERVED VIA VARYING COLOR DEVELOPMENT

The primary parameter predominantly studied in the nanobiotechnological syntheses of silver nanoparticles were the initial color transformation process, the progress and intensity in color development with time, after the addition of the plant extracts separately to $AgNO_3$ solutions. The color transformation of colorless and transparent silver nitrate solutions after the addition of aqueous plant extracts, from initial water color to orange brownish, indicated the reduction of Ag^+ to Ag^0 and the formation of silver nanoparticles. There were slight to drastic differences in the pace of color change and the final color developed of the different reaction mixtures (Figure 3.10), as a function of the varying quality of aqueous phytomass extracts added from varied plant materials. The different final color development observed in the reaction mixtures (color variation) is due to the different shapes of silver nanoparticles (rods, triangular, spherical, etc.), fabricated with the help of different plant extracts. The color development and the inference drawn from the observation are interpreted in Table 3.1. $AgNO_3$ solution and aqueous plant extracts individually showed no change in color with time.

3.3.2 COMPARATIVE ANALYSIS OF THE NANOPARTICLES THROUGH UV-VISIBLE SPECTRA

The UV-visible absorption spectra recorded from the nanoparticle suspensions of all the nine reaction mixtures, RM1, RM2, RM3, RM4, RM5, RM6, RM7, RM8 and RM9 after 48 h of reaction and complete stabilization with no further color transformation, are shown in Figures 3.11–3.17. In all the nine cases, a surface plasmon resonance (SPR) band absorption peak appeared between 410–480 nm, which is characteristic of silver nanoparticles [67]. The band peak positions of

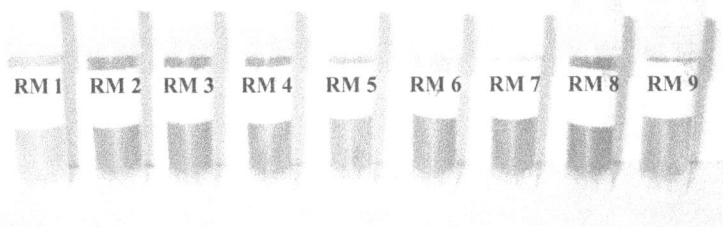

RM 1   RM 2   RM 3   RM 4   RM 5   RM 6   RM 7   RM 8   RM 9

FIGURE 3.10   Photograph of the reaction mixtures after complete stabilization.

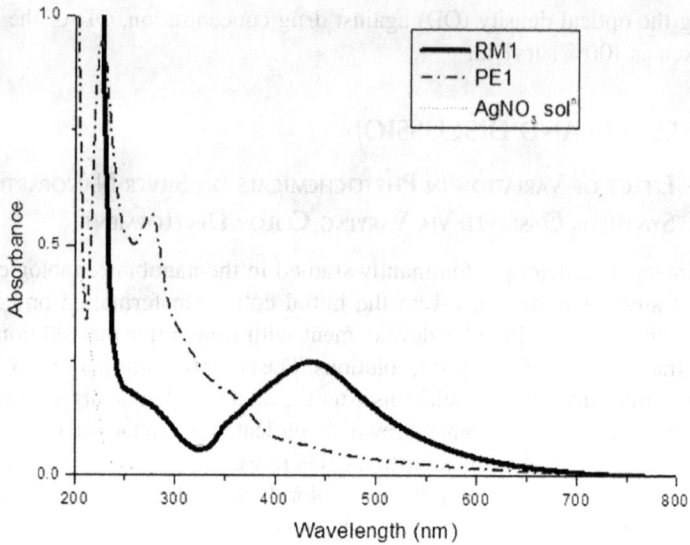

FIGURE 3.11 UV-visible spectra of silver nanoparticle suspension from reaction mixture RM1, Rudrakha extract PE1 and $AgNO_3$ only solution.

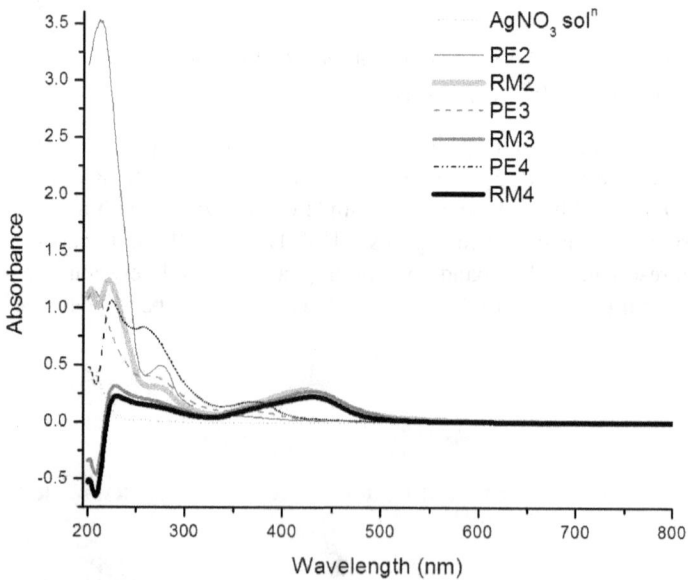

FIGURE 3.12 UV-visible spectra of silver nanoparticle suspension from reaction mixtures RM2, RM3, RM4; bark extract PE2, foliage extract PE3, fruit extract PE4 and $AgNO_3$ only solution.

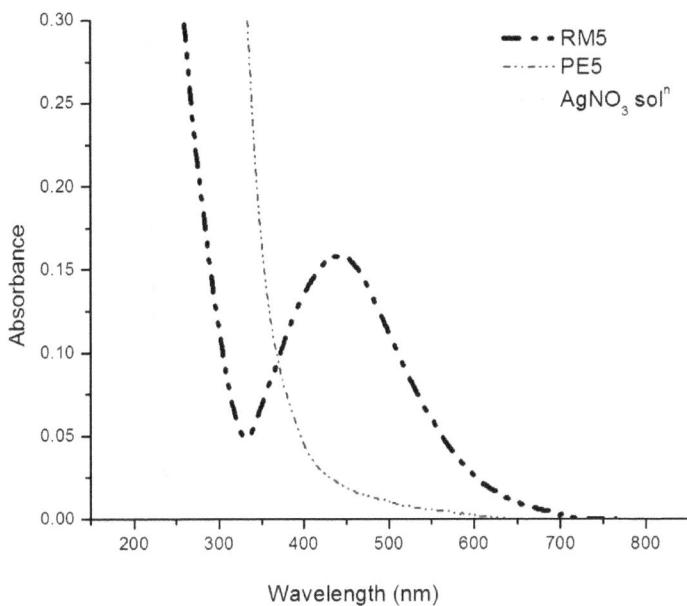

FIGURE 3.13 UV-visible spectra of silver nanoparticle suspension from reaction mixture RM5, foliage phytomass extract PE5 and $AgNO_3$ only solution.

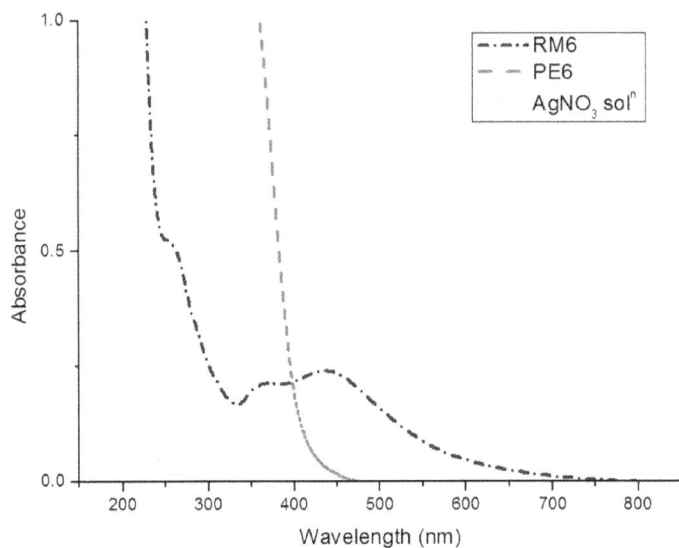

FIGURE 3.14 UV-visible spectra of silver nanoparticle suspension from reaction mixture RM6, foliage phytomass extract PE6 and $AgNO_3$ only solution.

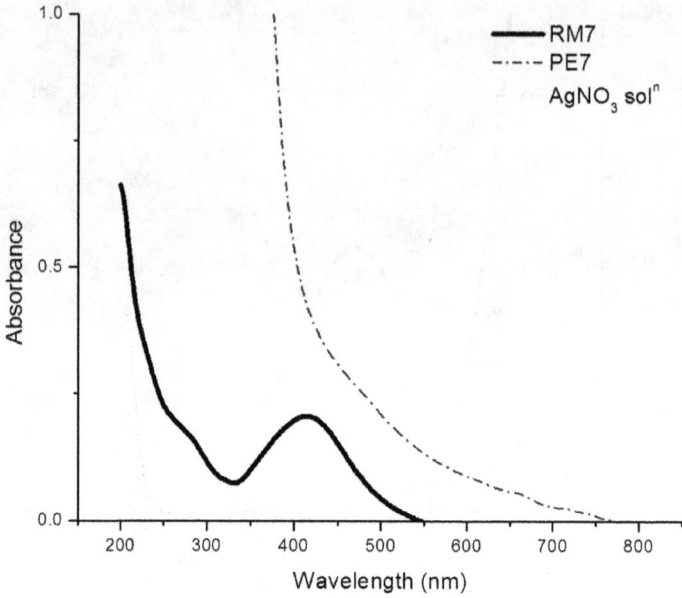

FIGURE 3.15 UV-visible spectra of silver nanoparticle suspension from reaction mixture RM7, foliage phytomass extract PE7 and AgNO$_3$ only solution.

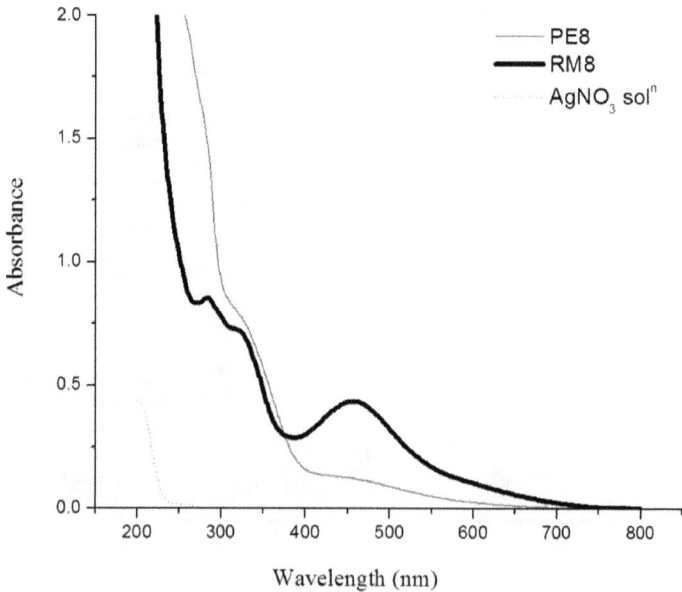

FIGURE 3.16 UV-visible spectra of silver nanoparticle suspension from reaction mixture RM8, foliage phytomass extract PE8 and AgNO$_3$ only solution.

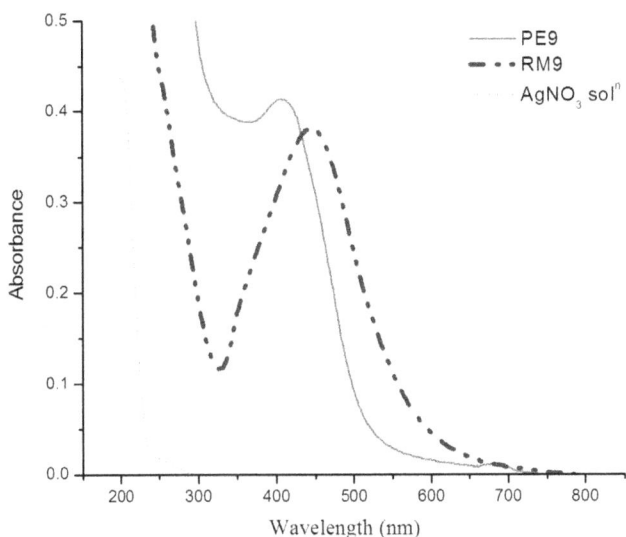

FIGURE 3.17 UV-visible spectra of silver nanoparticle suspension from reaction mixture RM9, rhizome phytomass extract PE9 and AgNO₃ only solution.

the nanoparticle suspensions from different reaction mixtures are systematically given in Table 3.2. The spectra of their corresponding plant extracts did not exhibit any similar noticeable peaks in this region. And also the absorption spectrum of only AgNO₃ solution existed rising at about 220 nm. According to the generalized theory [68], only a single SPR band peak is expected in the absorption spectra

TABLE 3.2

Band Peaks Obtained from UV-Visible Spectroscopy (Wavelength at Which Absorbance Is Maximum) of the Silver Nanoparticle Suspensions Generated through Diverse Phytomass Extracts

Nanoparticle Suspension from Reaction Mixture	Absorbance Peak (λ_{max} in nm)
RM1	438.0
RM2	421.9
RM3	430.4
RM4	428.5
RM5	440.5
RM6	436.5
RM7	415.0
RM8	455.5
RM9	445.0

of spherical nanoparticles, whereas anisotropic nanostructures or aggregates of spherical nanoparticles could give rise to two or more SPR bands depending on the shape of the particles.

3.3.3 TEM Observations and Qualitative Assessment of the Silver Nanoparticles

Morphological characteristics of the phytofabricated silver nanoparticles were observed through TEM. The ring-like diffraction pattern, exhibiting in all the cases studied, is indicative of crystalline nature of particles. This finding was reflected in the approximately circular nature of the selected area electron diffraction (SAED) spots displayed in Figures 3.18–3.26. While, Figures 3.27–3.35 show TEM images of silver nanoparticles formed, observations revealed that the silver nanoparticles were non uniform in size but dominantly spherical in shape belonging to suspensions of RM2, RM5 and RM7, while in other cases displayed a wide range of somewhat miscellaneous and irregular shapes. There was presence of interparticle interactions, which may have been due to the peripheral complexation of capped biomolecules. The silver nanostructures, generated through phytomass nanobiotechnology, showed wide variations in size distribution, in almost all of the reaction mixture suspensions, while high precision in the size of the nanoparticles with optimization was in the suspension from RM2.

3.3.4 Nanoparticle Size Determination

Details of the determination and analysis of the nanoparticles through particle size analyzer are vividly shown through the particle size distribution graph in

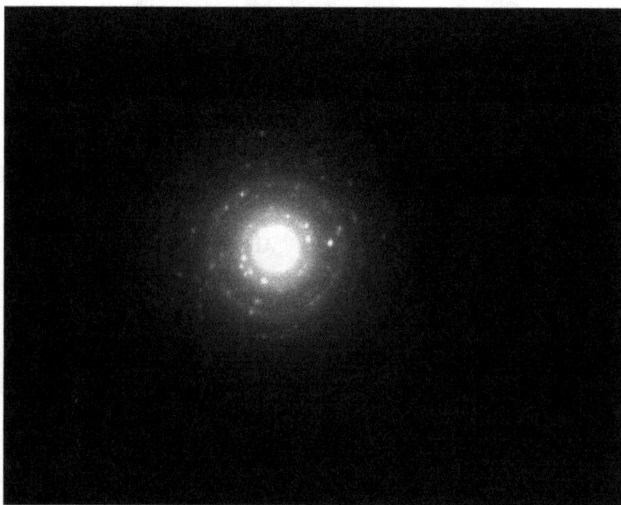

FIGURE 3.18   SAED pattern of silver nanoparticles from reaction mixture RM1.

FIGURE 3.19 SAED pattern of silver nanoparticles from reaction mixture RM2.

Figure 3.36. The size range of the majority of particles, developed through nanobiotechnology, complies with the nano range and lies in between 1 and 100 nm.

3.3.5 CHEMICAL COMPOSITION IDENTIFICATION

Chemical composition identification by EDAX gives evidence of the presence of silver in the nanoparticles from all the suspensions, but in variable quantity.

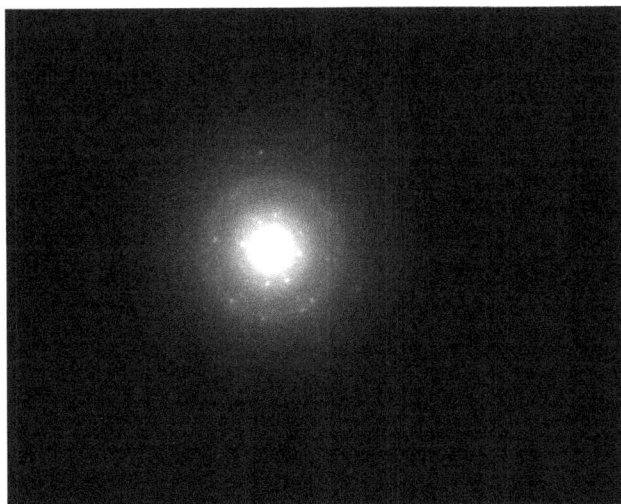

FIGURE 3.20 SAED pattern of silver nanoparticles from reaction mixture RM3.

FIGURE 3.21   SAED pattern of silver nanoparticles from reaction mixture RM4.

Details of the elemental composition of the nanoparticles formed from the diverse phytomass extracts for assessment are given in Table 3.3. The presence of elements other than silver in the spectra is due to possible interference of ions [69], during the biofabrication reaction by the varied phytochemicals. The participation of plant biomolecules in the reduction of Ag^+ to Ag^0 along with other

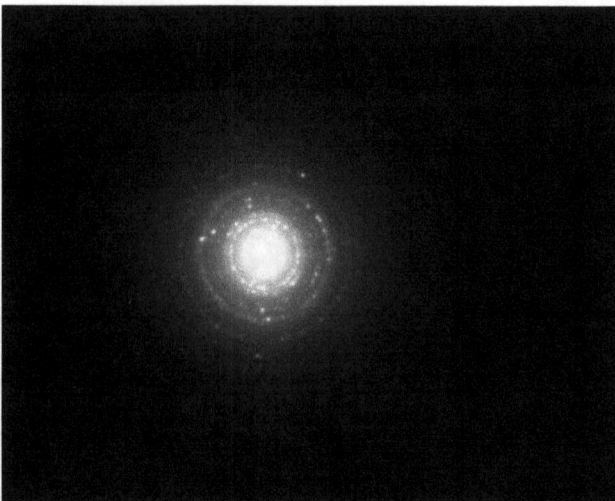

FIGURE 3.22   SAED pattern of silver nanoparticles from reaction mixture RM5.

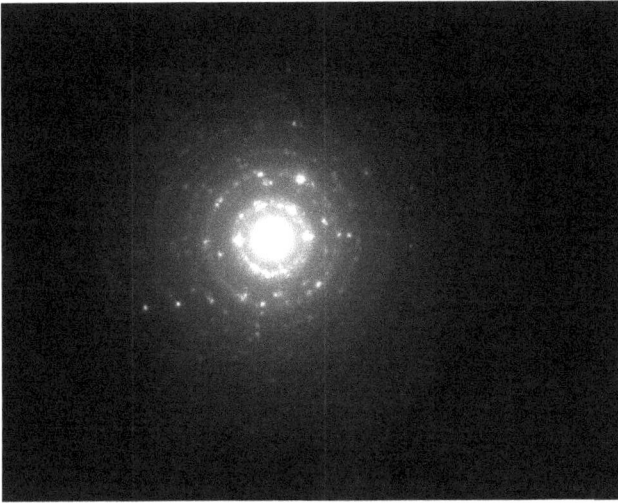

FIGURE 3.23 SAED pattern of silver nanoparticles from reaction mixture RM6.

functional groups capping the nanoparticles and acting as stabilizing ligands form an organic coat over the fabricated silver nanoparticles. This is the rationale behind the variable elemental composition of silver nanoparticles obtained from diverse phytomass extracts. The exposition has been further elucidated in the present study through FTIR spectra of the silver nanoparticles, which dealt with in the following chapters.

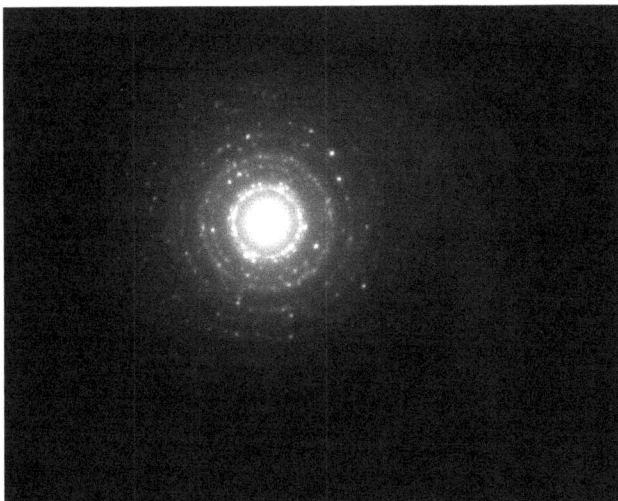

FIGURE 3.24 SAED pattern of silver nanoparticles from reaction mixture RM7.

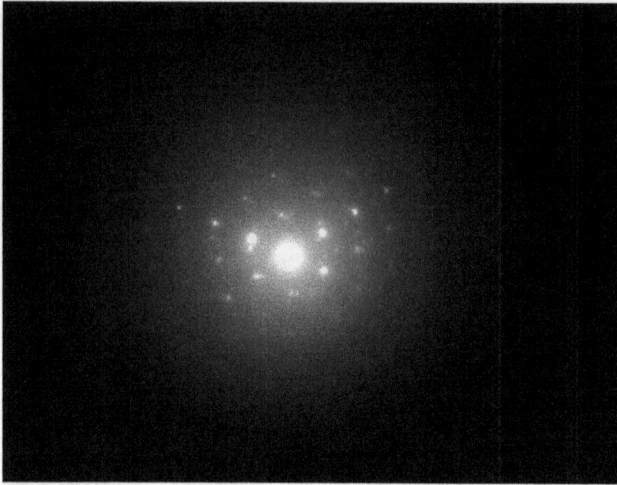

FIGURE 3.25 SAED pattern of silver nanoparticles from reaction mixture RM8.

3.3.6 ANTIMICROBIAL ASSAY

Silver nanoparticle residues obtained from the different reaction mixtures, such as RM1, RM2, RM3, RM4, RM5, RM6, RM7, RM8 and RM9, were assayed for antimicrobial activity against *S. aureus* (Gram positive) and *P. aeruginosa* (Gram negative) microbial strains, which cause majority of the infection despite aseptic measures and sterilization procedures in the biomedical sector. Disc diffusion protocol was adopted to determine the standard ZOI. Details of the results

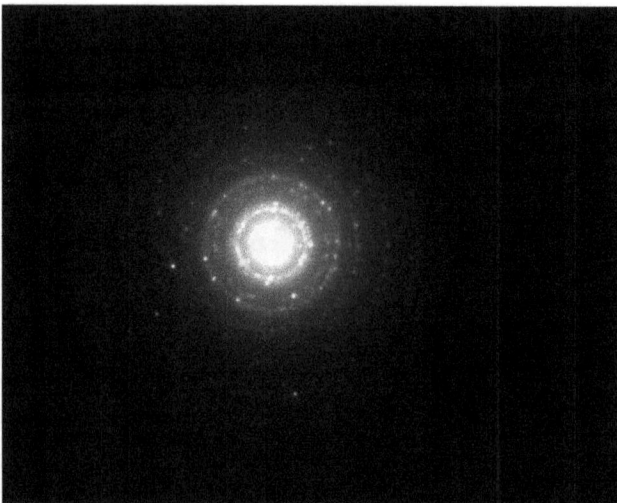

FIGURE 3.26 SAED pattern of silver nanoparticles from reaction mixture RM9.

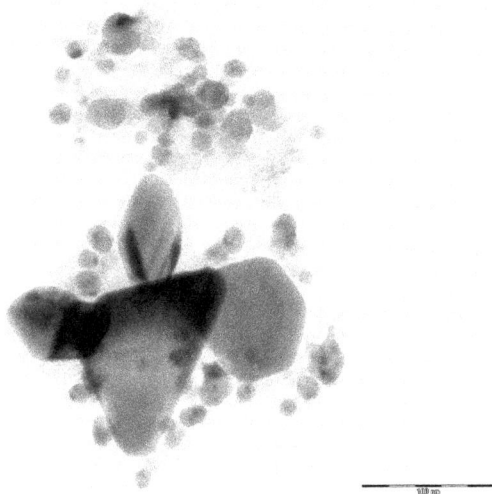

FIGURE 3.27 TEM micrograph of silver nanoparticles from reaction mixture RM1.

obtained are listed in Table 3.4. It has been observed in the present study that the effect was well pronounced against Gram-negative bacteria which contain only a thin peptidoglycan layer of 2~3 nm between the cytoplasmic membrane and the outer membrane, as well as against Gram-positive bacteria which lack the outer membrane but have a peptidoglycan layer of about 30-nm thickness. There exist several mechanisms behind the antimicrobial action of silver nanoparticles on bacteria, but are yet to be fully elucidated [70]. Out of them the most common are:

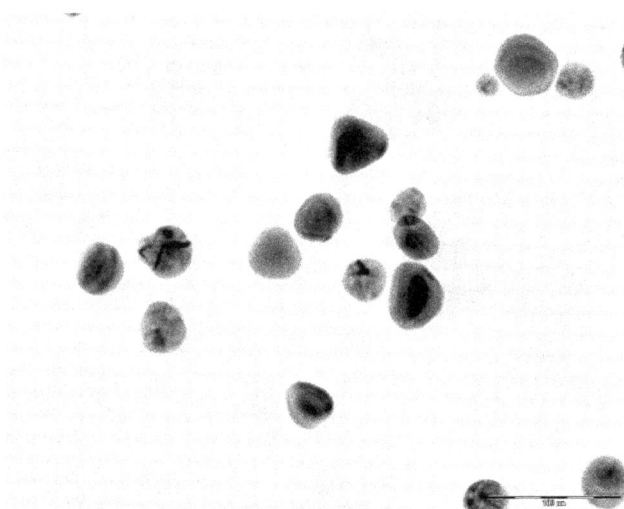

FIGURE 3.28 TEM micrograph of silver nanoparticles from reaction mixture RM2.

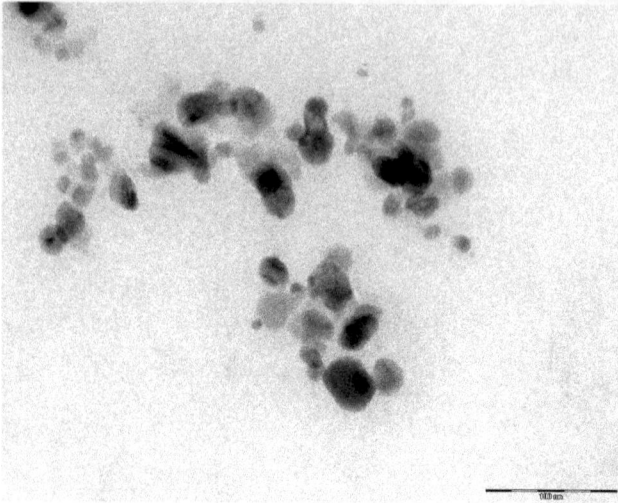

FIGURE 3.29   TEM micrograph of silver nanoparticles from reaction mixture RM3.

uptake of free Ag^+ released from the nanoparticles [71], followed by interruption in the production of ATP and replication of DNA [72], formation of some reactive oxygen species (ROS) [73], direct damage of cell membranes and causing its disruption [74].

Comparatively assessing the antimicrobial assay, the residue from RM1 was found to be most effective in inhibiting the microbes giving the highest and well pronounced ZOI against both the bacterial strains. The residue from RM6 also

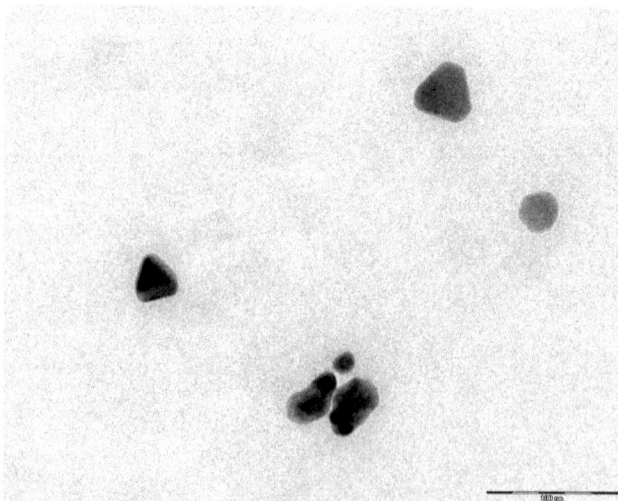

FIGURE 3.30   TEM micrograph of silver nanoparticles from reaction mixture RM4.

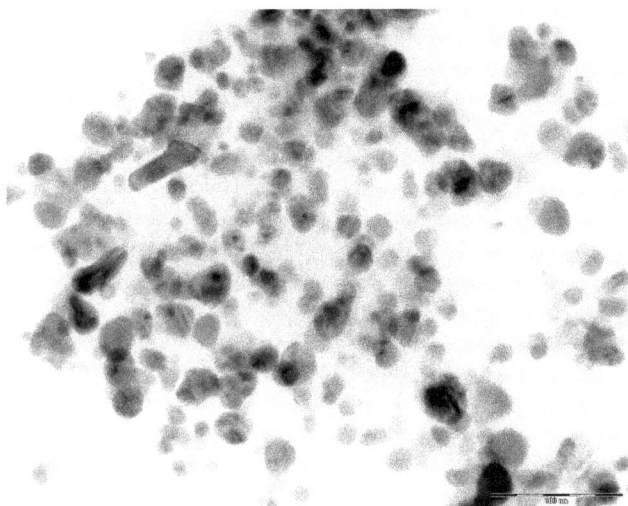

FIGURE 3.31   TEM micrograph of silver nanoparticles from reaction mixture RM5.

gave remarkable ZOI. For the other testing agents, control positive (C+), ZOI were well recorded, while for the control negative (C−), ZOI were not formed against any of the bacterial strains. Therefore, from our present observation, we have perceived that there are several key factors responsible for overwhelming microbial inhibition; the size distribution of nanosilver definitely playing a vital role, while the chemical composition of the nanoparticles together with the percentage and

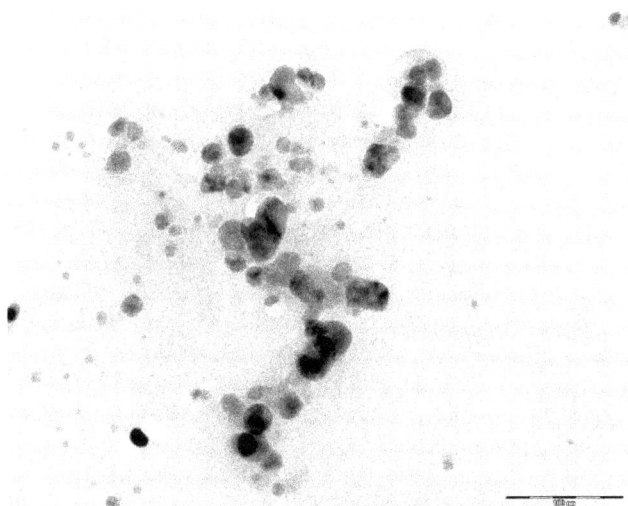

FIGURE 3.32   TEM micrograph of silver nanoparticles from reaction mixture RM6.

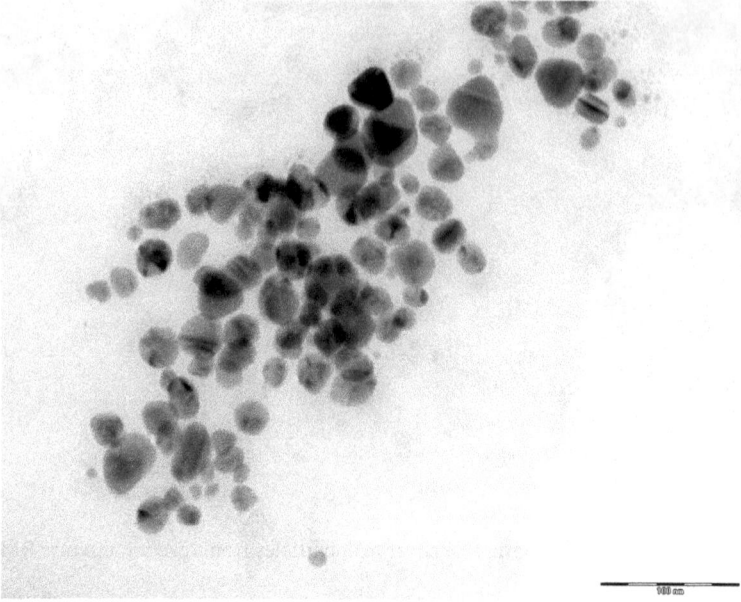

FIGURE 3.33   TEM micrograph of silver nanoparticles from reaction mixture RM7.

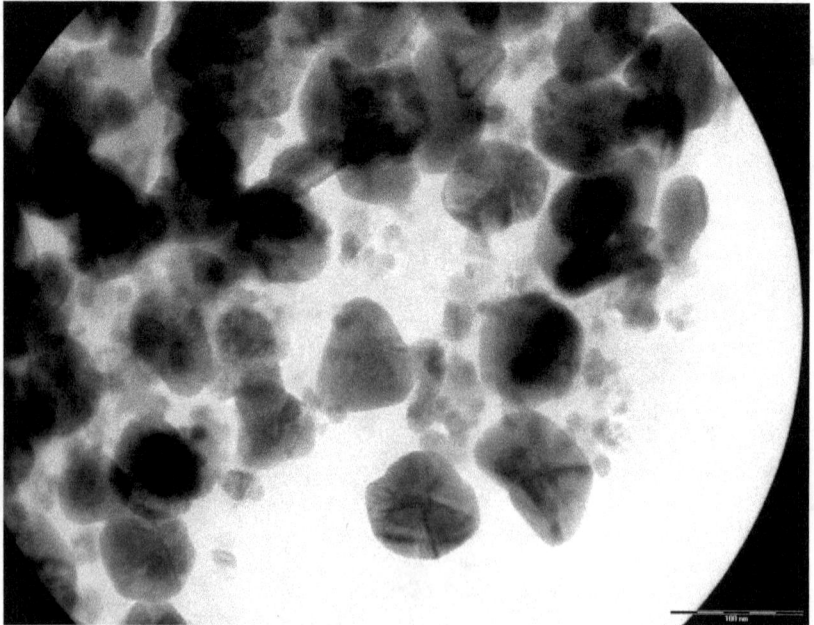

FIGURE 3.34   TEM micrograph of silver nanoparticles from reaction mixture RM8.

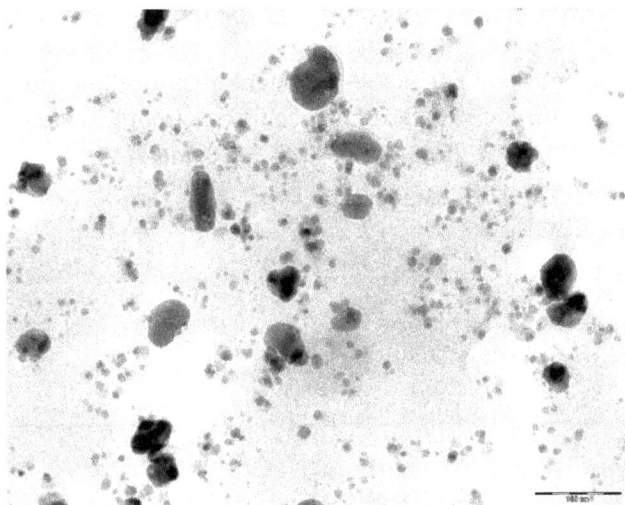

FIGURE 3.35   TEM micrograph of silver nanoparticles from reaction mixture RM9.

presence of high silver content is also crucial [19]. The bactericidal activity of the silver nanoparticles depends on their stability in the cultured medium too.

3.3.7   CYTOTOXICITY TEST

The silver nanoparticles synthesized can be considered reasonably biocompatible and safe to be used in regulated quantity based on the cytotoxicity test done. The CC_{50} value of the silver nanoparticle suspensions against the J774A.1 murine

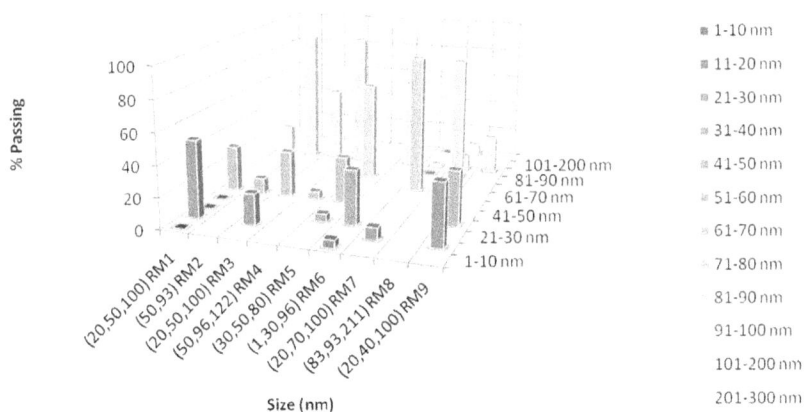

FIGURE 3.36   Histograms of particle size distribution of nanoparticles obtained from different reaction mixtures.

TABLE 3.3

Chemical Composition Elucidation: Element Identification and Quantification through EDAX, of Nanoparticles from Reaction Mixtures RM1, RM2, RM3, RM4, RM5, RM6, RM7, RM8 and RM9

Element		RM1	RM2	RM3	RM4	RM5	RM6	RM7	RM8	RM9
Carbon	Weight%	16.5	57.1	27.7	11.1	26.6	27.1	72.6	43.2	29.4
CK	Atomic%	39.4	77.3	42.9	16.2	33.3	45.4	78.7	69.4	46.7
Oxygen	Weight%	20.4	16.9	14.1	53.3	70.5	15.6	26	19.9	40.2
OK	Atomic%	36.4	17.2	33.5	58.6	66.3	36	21.2	24.0	47.9
Silver	Weight%	55.4	20.5	51.6	03.4	03	49.4	01.4	36.9	30.4
AgL	Atomic%	14.7	03.1	10.6	00.6	00.4	9.9	00.2	06.6	05.4

TABLE 3.4

Zone of Inhibition (ZOI) in (mm) against Selective Microbial Strains, *Staphylococcus aureus* and *Pseudomonas aeruginosa*, by Silver Nanoparticle Pellets Phytonanofabricated Involving Nanobiotechnology from Different Reaction Mixtures and ZOI by the Controls

Name of the Micro-organism	RM1	RM2	RM3	RM4	RM5	RM6	RM7	RM8	RM9	Sparfloxacin (C+)	H$_2$O (C−)
Staphylococcus aureus	21	18	19	17	18	21	14.3	16.4	18.4	41.5	0
Pseudomonas aeruginosa	25	19.4	20.3	18.2	18.4	25	16.4	16.4	19	42.5	0

macrophage cell line, obtained were the following, RM5 (0.282 ± 0.0006 µg/mL), RM7 (0.388 ± 0.01 µg/mL) while that of RM9 (0.256 ± 0.003 µg/mL). These data represent the mean ± SD of experiments performed in triplicate. Results of the analysis are given in Table 3.5, and graphs of the optical density plotted against concentration in Figures 3.37–3.39.

TABLE 3.5

Cytotoxicity Analysis of Silver Nanoparticles

Sample	CC$_{50}$	SD
RM5	0.2823036 µg/mL	0.0006442
RM7	0.3882364 µg/mL	0.0128339
RM9	0.2562295 µg/mL	0.0037729

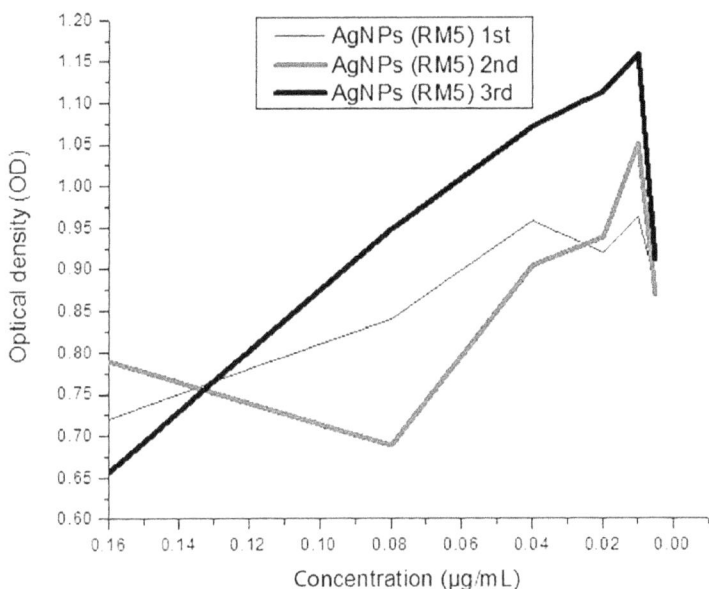

FIGURE 3.37 Graph of the optical density plotted against concentration for silver nanoparticles from suspension RM5, performed in triplicates.

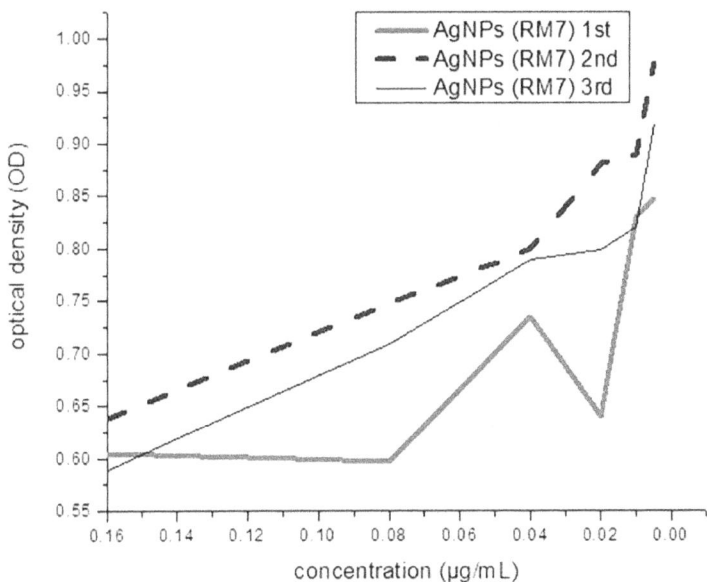

FIGURE 3.38 Graph of the optical density plotted against concentration for silver nanoparticles from suspension RM7, performed in triplicates.

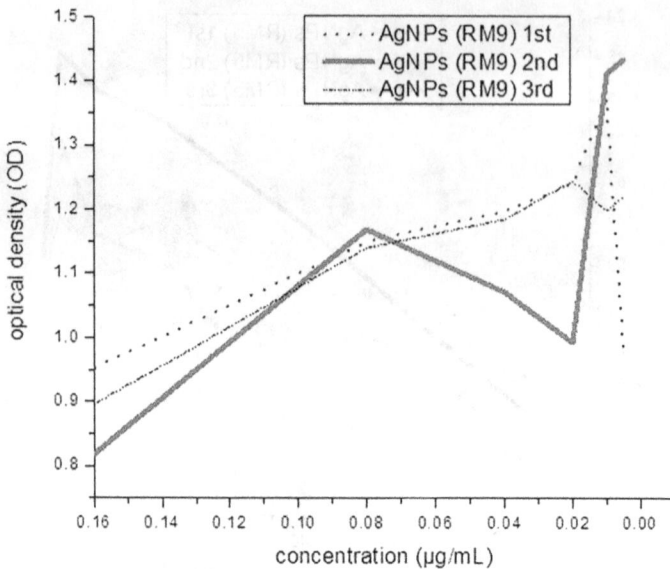

FIGURE 3.39 Graph of the optical density plotted against concentration for silver nanoparticles from suspension RM9, performed in triplicates.

3.4 CONCLUSION

This chapter emphasizes that the intervention of the green herbal route for nanosilver synthesis is a convenient, less time-consuming approach and suitable for large-scale production. This method is also a safe-synthetic protocol, while discarding conventional methods and toxic reagents that open the way toward physiologically and ecologically harmful by-products. Obnoxious reagents used for syntheses, persistently adhere to the surface of the formed nanostructures rendering them extremely harsh to be handled and to be applied, whereas in this aforementioned case studies the beneficiaries of the herbal phytomass are adhering to the nanobiotechnologically synthesized silver nanoparticles.

Marching into the upcoming phase of the ameliorated face having herbalism merged with nanotechnology in the form of nanobiotechnology, the inference drawn therefore precisely states as well as the statistics vividly elucidates that the dried fruit of *Elaeocarpus ganitrus* Roxb., i.e. the Rudraksha bead synthesizes silver nanoparticles gradually but most efficiently, effectively showing an electrifying eminence with an elegant prospect in this emerging field. The prominent feature of the Rudraksha bead is its reuse potentiality and non-degradability of the plant material after preparation of the aqueous extract. Therefore, having the inexhaustible capability to produce silver nanoparticles prolifically, mediating a number of repeated syntheses with the same bead, without any noticeable disintegration and degradation. Probably the prolonged reuse of Rudraksha bead for the preparation of plant extract is due to its lignified woody and stony endocarp which delays degradation and perish. The Rudraksha seeds enclosed within the endocarp are rich store house of plant metabolites, which are only partially released into the aqueous solution on preparation of water extract and conduct the bioreduction process of nanoparticle phytofabrication. It's this high content of bioorganic molecules together with the noteworthy non-consumable property and non-degradability demarcates it from other perishable plant materials.

Through this study, it can also be concluded that the physics and chemistry of the silver nanoparticles produced, using the phytofabrication technique, can be controlled through tailored adjustments. Obtained data indicates that the quality is consistent with the quantity of phytomass mediating distinctively matters for the pace of proceeding biogenesis; shape, size and size distribution along with other physico-chemical properties of the nanoparticles generated. These properties can possibly be tuned via the variable parameters and the several factors governing the experimental conditions. Based on the shape and size, the nanoparticles also exhibit different properties (antimicrobial, optical, etc.). Furthermore, development of nanobiomaterials and bio-inspired materials through phytomass via nanobiotechnology, i.e. phytonanofabrication of nanoparticles, can prove to be the cutting edge of research in modern nanotechnology.

REFERENCES

1. P. Dwivedi, D. Tiwary, P. K. Mishra and J. P. Chakraborty, Adv. Sci. Eng. Med., 12 (2020) 548–555.
2. P. Dwivedi, D. Tiwary, P. K. Mishra and J. P. Chakraborty, Nano-Struct. Nano-Objects, 22 (2020) 100485, 1–7.
3. P. Dwivedi, D. Tiwary, P. K. Mishra, S. S. Narvi and R. P. Tewari, Inorg. Chem. Commun., 126 (2021) 108479, 1–12.
4. P. Dwivedi, S. S. Narvi and R. P. Tewari, Int. J. Green Nanotechnol., 4 (2012) 248–261.
5. P. Dwivedi, S. S. Narvi and R. P. Tewari, Adv. Mat. Res., 585 (2012) 144–148.
6. P. Dwivedi, S. S. Narvi and R. P. Tewari, Silver Nanoparticles and Nanocomposites. Encycl. Biomed. Polym. Polym. Biomater, 10 (2015) 7275–7285.
7. P. Dwivedi, S. S. Narvi and R. P. Tewari, Int. J. Biomed. Nanosci. Nanotechnol., 2 (2012) 187–206.
8. P. Dwivedi, S. S. Narvi and R. P. Tewari, J. Chin. Med. Res. Develop., 1 (2012) 23–27.
9. P. Dwivedi, S. S. Narvi and R. P. Tewari, Int. J. Adv. Eng. Sci. Technol., 2 (2012) 236–243.
10. P. Dwivedi, S. S. Narvi and R. P. Tewari, Int. J. Eng. Res. Appl., 2 (2012) 1490–1495.
11. P. Dwivedi, S. S. Narvi and R. P. Tewari, Int. J. Sci. Res. Publ., 2 (2012) 1–5.
12. P. Dwivedi, S. S. Narvi and R. P. Tewari, Ann. Res. Rev. Biol., 4 (2014) 1059–1069.
13. P. Dwivedi, S. S. Narvi and R. P. Tewari, Adv. Sci. Eng. Med., 6 (2014) 1–9.
14. P. Dwivedi, S. S. Narvi and R. P. Tewari, Nano LIFE, 5 (2015) 1540006.
15. C. L. Fox, Arch. Surg., 96 (1968) 184–188.
16. P. Dwivedi, D. Tiwary, S. S. Narvi, R. P. Tewari and K. P. Shukla, Lett. Appl. Nanobiosci., 9 (2020) 1485–1493.
17. N. Stobie, B. Duffy, D. E. McCormack, J. Colreavy, M. Hidalgo, P. McHale, et al., Biomaterials, 29 (2008) 963–969.

18. D. R. Monteiro, L. F. Gorupb, A. S. Takamiyaa, A. C. Ruvollo-Filho, E. R. de Camargob and D. B. Barbosa, Int. J. Antimicrob. Agents, 34 (2009) 103–110.
19. P. Dwivedi, S. S. Narvi and R. P. Tewari, Ind. Crops Prod., 54 (2014) 22–31.
20. R. K. Singh, S. K. Bhattacharya and S. B. Acharya, Phytomedicine, 7 (2000) 205–207.
21. R. K. Singh and B. L. Pandey, J. Aromat. Plant Sci., 21 (1999) 1030–1032.
22. R. K. Singh and G. Nath, Phytother. Res., 13 (1999) 448–450.
23. V. B. Panday and S. K. Bhattacharya, J. Res. Edu. Ind. Med., 4 (1985) 47–50.
24. P. Rashmi and K. Amrinder, Int. J. Res., 1 (2014) 334–353.
25. Z. M. H. Khan, H. M. Faruquee and M. M. Shaik, Med. Plant Res., 3 (2013) 70–77.
26. S. J. Padayatty, A. Katz, Y. Wang, P. Eck, O. Kwon, J. H. Lee, S. Chen, C. Corpe, A. Dutta, S. K. Dutta and M. Levine, J. Am. Coll. Nutr., 22 (2003) 18–35.
27. S. Malik, S. Mann, D. Gupta and R. K. Gupta, J. Pharmacogn. Phytochem., 2 (2013) 66–73.
28. A. Kaur, A. C. Rana, V. Tiwari, R. Sharma and S. Kumar, J. Appl. Pharm. Sci., 1 (2011) 6–11.
29. A. N. M. Mamun-or-Rashid, M. Azam, B. K. Dash, F. B. Hafiz and M. K. Sen, Mintage J. Pharm. Med. Sci., 2 (2013) 37–42.
30. M. Akram, U. Shahab, A. Afzal, U. Khan, H. Abdul, E. Mohiuddin and M. Asif, Res. J. Biol. Plant. Biol., 55 (2010) 65–70.
31. N. R. Farnsworth, J. Pharm. Sci., 55 (1966) 225–286.
32. S. S. Sakat, S. S. Wankhede, A. R. Juvekar, V. R. Mali and S. L. Bodhankar, Int. J. Pharm. Tech. Res., 1 (2009) 779–782.
33. B. Singh, A. Chopra, M. P. S. Ishar, A. Sharma and T. Raj, Indian J. Pharm. Sci., 72 (2010) 261–265.
34. B. Singh, V. Sharma, M. P. S. Ishar and A. Sharma, Nat. Prod. J., 3 (2013) 224–229.
35. J. D. Kirtikar and B. D. Basu, Indian Medicinal Plants, L. M. Basu(Publisher), Allahabad, India, 9 (1994) 1023, 1774, 1784.
36. M. C. Sharma and S. Sharma, Int. J. Microbiol. Res., 1 (2010) 166–170.
37. L. Oberoi, T. Akiyama, K. H. Lee and S. J. Liu, Phytomedicine, 18 (2011) 259–265.
38. S. A. S. Chatha, A. I. Hussain, R. Asad, M. Majeed and N. Aslam, J. Food. Process Technol., 5 (2014) 1–5.
39. K. Gopinath, S. Gowri, V. Karthika and A. Arumugam, J. Nanostruct. Chem., 4 (2014) 1–11.
40. M. B. Davies, J. Austin and D. A. Partridge, Vitamin C: Its Chemistry and Biochemistry, The Royal Society of Chemistry, ISBN 0–85186–333–7 (1991) 48.
41. J. D. Horner, J. Chem. Ecol., 14 (1988) 1227–1237.
42. B. Winkel-Shirley, Plant Physiol., 126 (2001) 485–493.
43. A. Gupta, G. Sharma, S. Pandey, B. Verma, V. Pal and S. S. Agrawal, Int. J. Pharm. Sci. Rev. Res., 27 (2014) 328–333.
44. N. Chowdhary, M. Kaur, A. Singh and B. Kumar, IJRPB., 2 (2014) 1071–1081.
45. P. Pattanayak, P. Behera, D. Das and S. K. Panda, Phcog. Rev., 4 (2010) 95–105.
46. L. Srinivas, V. K. Shalini and M. Shylaja, Arch. Biochem. Biophys., 292 (1992) 617–623.
47. K. P. V. Subbaiah, G. Ramanjaneyulu and N. Savithramma, ISRJ., 3 (2013) 1–7.
48. A. Maheswaran, P. Brindha, A. R. Mullaicharam and V. Ravichandiran, Am. J. Pharm. Tech. Res., 4 (2014) 905–911.
49. R. Jagannathan, P. M. Abraham and P. Poddar, J. Phys. Chem. B., 116 (2012) 14533–14540.
50. E. Janata, A. Henglein and B. G. Ershov, J. Phys. Chem., 98 (1994) 10888–10890.

51. M. Chen, Y. G. Feng, X. Wang, T. C. Li, J. Y. Zhang and D. J. Qian, Langmuir, 23 (2007) 5296–5304.
52. S. S. Shankar, A. Ahmad, R. Pasricha and M. J. Sastry, Mater. Chem., 13 (2003) 1822–1846.
53. S. S. Shankar, A. Rai, A. Ahmad and M. Sastry, J. Colloid. Interface. Sci., 275 (2004) 496–502.
54. A. R. Vilchis-Nestor, V. Sanchez-Mendieta, M. A. Camacho- Lopez, R. M. Gomez-Espinosa and J. A. Arenas-Alatorre, Mater. Lett., 62 (2008) 3103–3105.
55. J. Y. Song, E. Y. Kwon and B. S. Kim, Bioprocess. Biosyst. Eng., 33 (2010) 159–164.
56. A. Singh, M. Talat, D. Singh and O. N. Srivastava, J. Nanoparticle Res., 12 (2010) 1667–1675.
57. N. Ahmad, S. Sharma, M. K. Alam, V. N. Singh, S. F. Shamsi, B. R. Mehta and A. Fatma, Coll. Surf. B. Biointerfaces, 81 (2010) 81–86.
58. M. F. Zayed, W. H. Eisa and A. A. Shabaka, Spectrochim. Acta. A. Mol. Biomol. Spectrosc., 98 (2012) 423–428.
59. L. C. Gruen, Biochim. Biophys. Acta., 386 (1975) 270–274.
60. Y. N. Tan, J. Y. Lee and D. I. Wang, J. Am. Chem. Soc., 132 (2010) 5677–5686.
61. J. Glusker, A. Katz and C. Bock, Rigaku J., 16 (1999) 8–16.
62. S. Panigrahi, S. Kundu, S. Ghosh, S. Nath and T. Pal, J. Nanoparticle Res., 6 (2004) 411–414.
63. H. Korbekandi, Z. Ashari, S. Iravani and S. Abbasi, Iran J. Pharm. Res., 12 (2013) 289–298.
64. N. Ahmad, S. Sharma, V. N. Singh, S. F. Shamsi, A. Fatma and B. R. Mehta, Biotechnol. Res. Int., Article ID 454090 (2011) 1–8.
65. S. Si and T. K. Mandal, Tryptophan-based Peptides to Synthesize Gold and Silver Nanoparticles: A Mechanistic and Kinetic Study, Chemistry, 13 (2007) 3160–3168.
66. V. V. Makarov, A. J. Love, O. V. Sinitsyna, S. S. Makarova, I. V. Yaminsky, M. E. Taliansky and N. O. Kalinina, Acta Naturae, 6 (2014) 35–44.
67. P. Mulvaney, Surface Plasmon Spectroscopy of Nanosized Metal Particles, Langmuir, 12 (1996) 788–800.
68. R. Das, S. S. Nath, D. Chakdar, G. Gope and R. Bhattacharjee, J. Exp. Nanosci., 5 (2010) 357–362.
69. A. Hanglein, Chem. Mater., 10 (1998) 444–450.
70. B. Lansdown, J. Wound Care, 11 (2002) 125–130.
71. P. Dibrov, J. Dzioba, K. K. Gosink and C. C. Hase, Antimicrob. Agents Chemother., 46 (2002) 2668.
72. C. Lok, C. Ho, R. Chen, Q. He, W. Yu, H. Sun, et al., J. Proteome. Res., 5 (2006) 916–924.
73. H. Park, J. Kim, J. Kim, J. Lee, J. Hahn, M. Gu, et al., Water Res., 43 (2009) 1027–1032.
74. M. Raffi, F. Hussain, T. Bhatti, J. Akhter, A. Hameed and M. Hasan, J. Mater. Sci. Technol., 24 (2008) 192–196.

4 Biocompatible Bioactive Silver/Chitosan (Ag/CS) Nanocomposite Surface Coating System
Nanobiotechnological Synthesis

4.1 INTRODUCTION

As the progress in nanoscience is marching ahead with time, having surpassing applications knocking out challenging problems in almost every field, the set back of biomaterials associated infections (BAI) is still procrastinating in the biomedical arena. The race prevailing between tissue integration and microbial adhesion becomes a major perioperative cause of concern during the biomedical implantation process. Microbial adhesion further gives rise to biofilm formation which finally leads to implant failure. There are reports of failure of such devices stems from bacterial biofilm build up [1, 2], which is extremely resistant to host defense mechanisms and antibiotic treatment [3]. Often the only solution which remains to an infected implanted device is its surgical removal.

Surging ahead researches in this direction has found that the most commonly reported pathogen causing nosocomial and hospital-acquired infections even in intensive care unit patients is *Staphylococcus* [4–6]. Most important in the pathogenesis of foreign-body associated infections is the ability of these bacteria to colonize the biomaterial surface by the formation of a thick, multilayered biofilm [7]. Generally, bacterial adhesion to biomaterial surfaces is the first essential step in the pathogenesis of these infections [8, 9]. Small numbers of bacteria from the patient's skin or mucous membranes, where these bacteria normally thrive, probably contaminate the biomaterial during the surgical implantation process. The bacteria very frequently acquired through the hands of the clinical or surgical staffs, from contaminated hospital environment, other patients or can reach from the distant local infections. When the bacteria rapidly adhere to the biomaterial, they start to proliferate to form multilayered cell clusters and gradually there is biofilm formation on the biomaterial surface [10].

DOI: 10.1201/9781003217343-4

Bacterial adhesion to the surfaces comprises of the initial attraction of the bacterial cells to the surfaces followed by the adsorption process and then attachment [11]. Both specific together with non-specific interactions demonstrate a pivotal role in capacitating the bacterial cell to attach to or even to detach from the adhered biomaterial surface [12]. The relative contributions of specific and non-specific mechanisms are chiefly depended on the surface properties of the biomaterial. Moreover, bacterial adhesion is an extremely complicated process that is affected and influenced by many factors; primarily including the material surface characteristics, especially its physico-chemical properties, topography, surface roughness, physical configuration, chemical composition, surface charge or hydrophobicity [13–17]. Increase in the hydrophobicity of the surfaces enhances the probable adherence of prevalent hydrophobic bacteria to the surfaces such as of the biomaterials and henceforth their clustering.

Therefore, in this chapter there is an attempt to design a biocompatible, but bioactive, i.e. self-sterilizing having high antimicrobial properties, and biodegradable silver/chitosan (Ag/CS) nanocomposite being also termed as Ag/CS bionanocomposite, material. The nanomaterial simultaneously possessed with optimum hydrophilic characteristic can be applied for coating of biomaterials, such as medical implants and surgical devices. This nanomaterial will impart surface modification to imply hindrance to microbial adhesion and thus resistance to BAI.

Ongoing researches in the biomedical arena have brought forward the hypothesis that silver-containing material can minimize implant associated nosocomial infection which have been confirmed by several in vitro and in vivo experiments [18–22]. Silver nanoparticles having high surface reactivity due to high surface to volume ratio will release silver ions (Ag^+) which are antimicrobial in nature. These Ag^+ ions have the ability to kill a very broad spectrum of medically relevant bacteria (Gram positive and Gram negative) as well as fungi (molds and yeasts). Ionic silver is also oligodynamic, which means that it is antimicrobial even at very low doses, as low as about 0.001–0.05 ppm. Although silver is a heavy metal, at the reference low concentration level, it is nontoxic to human cells and therefore very safe.

CS is used as the matrix, which can protect silver nanoparticles from uncontrolled oxidation and stabilize them from agglomeration. CS is a natural biopolymer derived by partial deacetylation of chitin, a component commonly found in the exoskeleton of some crustaceans, and is the second most abundant biopolymer after cellulose. This biopolymer is composed of poly (β-[1-4]-2-amino-2-deoxy-D-glucopyranose) and has many advantageous features like biocompatibility, biodegradability, non-toxicity, hydrophilicity, together with antimicrobial properties and is approved by *Food and Drug Administration* (FDA) [23, 24].

It is for the first time here when silver nanoparticles have been biosynthesized using foliage needles of the plant *Pseudotsuga menziesii* (the Christmas tree); dispersed in CS, a biopolymer matrix, thereby the bionanocomposite, self-sterilizing coating biomaterial, i.e. the nanobiomaterial has been fabricated. This nanobiomaterial, designed for coating of medical implants, when leached out will cause minimal harm to the human body. The efficacy of the bioactive, *i.e.* self-sterilizing, biocompatible and biodegradable Ag/CS nanocomposite (or the Ag/CS bionanocomposite), nanobiomaterial against *Staphylococcus aureus* biofilm has also been studied here, after being coated over medical grade implant.

4.2   EXPERIMENTAL SECTION

4.2.1   MATERIALS

CS (degree of deacetylation: 79%, molecular mass: 500,000 g/mol) was purchased from Sea Foods (Cochin), India; acetic acid glacial (extra pure), acetone, glutaraldehyde (GA), Muller Hinton agar and nutrient broth from Thomas Baker (Chemical) Pvt. Ltd. India. Silver nanoparticle suspension from reaction mixture (R5), phytofabricated by the foliage needles of the plant *Pseudotsuga menziesii*. Bacterial strains *Pseudomonas aeruginosa* (Gram negative), and *S. aureus* (Gram positive) were obtained from the culture bank of Microbiology Department, Sam Higginbottom Institute of Agriculture, Technology and Sciences, Allahabad, India. Deionized water was used as a solvent for preparing the solutions.

4.2.2   SYNTHESIS – PHYTOMASS SYNTHESIS OF AG/CS NANOCOMPOSITE

The silver nanoparticles obtained from the reaction mixture (RM5), i.e. phytofabricated by the foliage needles of *Pseudotsuga menziesii* (experimental details referred and discussed in Chapter 3 of this monograph), was selectively taken. This was done on the basis of its impressive result, together with the high vitamin C content and astringency present in the foliage of the plant, having very good wound healing and anti-inflammatory property, respectively. These silver nanoparticles from reaction mixture (RM5) and the residue denoted in this chapter as (R5) have been fabricated by bottom-up approach through their self-assembly and colloidal aggregation, after phytochemicals-mediated bio-reduction of $Ag^+ \rightarrow Ag^0$. Plant extracts from the needles of *Pseudotsuga menziesii* (the Christmas tree) have been used for this purpose, as discussed in Chapter 3. When complete bio-reduction of Ag^+ ions and stabilization of nanoparticles was confirmed through no noticeable color change, the reaction mixture (RM5) was centrifuged at 9,000 rpm for 15 min and the residue of silver nanoparticles were re-dispersed in distilled water. This procedure was repeated three times to isolate and purify the nanoparticles from other unbound bio-organic molecules also present in the suspension. The remnant residue was thereafter dispersed in 15 mL of CS solution (2% [w/v] in 1% [v/v] acetic acid) in the ratio of 1:1 and the dispersion was sonicated for 10 min. Finally, the biocompatible and biodegradable Ag/CS nanocomposite (Ag/CS bionanocomposite) solution was developed, which was used for coating of medical implants and for thin film preparation.

4.2.3 COATING AND DEVELOPMENT OF THIN FILM

Coating of medical implants, such as stainless steel screw, was done by dipping the implants in the bionanocomposite material of silver reinforced CS matrix (dip coating), and then by air drying. Stainless steel rod and stainless steel screw were also coated by pouring the bionanocomposite material over the implants (solvent casting technique) and air dried. Facile and less time consuming methods of dip coating and solvent casting were adapted for preliminary and basic studies [25].

Thin film was also prepared separately by casting the Ag/CS bionanocomposite solution on glass slab, air drying at room temperature with ambient conditions for 48 h and peeling off thereafter [26]. The thickness of the coating was measured, and scratch test of the coated surface was also done along with several other characterizations of the Ag/CS bionanocomposite.

4.2.4 FTIR SPECTROSCOPY

4.2.4.1 FTIR Study of the Silver Nanoparticles

The dried residue of silver nanoparticles, obtained from the reaction mixture (RM5), phytofabricated by the foliage needles of *Pseudotsuga menziesii*, labeled for this experimental study as (R5), was characterized using Fourier transformed infrared spectroscopy (FTIR). The FTIR analysis was carried out to have a general idea of the possible biomolecules responsible for the reduction of Ag^+ ions and identify the functional groups capping and efficiently stabilizing the silver nanoparticles phytofabricated by the plant extract. This also provides a brief account of the probable beneficiaries of the plant adhered to the nanoparticles. The isolated and purified residue of silver nanoparticles from other unbound bio-organic compounds present in the suspension was received after repeated centrifugation. FTIR spectrum of these silver nanoparticles was recorded over the range of (400–4000) cm^{-1} with (PerkinElmer) spectrophotometer for studying the chemical properties.

4.2.4.2 FTIR Study of the Ag/CS Bionanocomposite Film

FTIR spectra were also recorded over the range of (400–4000) cm^{-1} with the spectrophotometer (FTLA 2000 ABB) to assess the chemical properties of only CS and Ag/CS bionanocomposite films.

4.2.5 MORPHOLOGICAL STUDY AND CHARACTERIZATION
OF THE AG/CS NANOCOMPOSITE

The Ag/CS bionanocomposite film was coated with a thin layer of graphite and examined in scanning electron microscope (SEM) using (QUANTA200, FEI Ltd.) and (JEOL JXA 8100) instruments for studying the morphological nature. The surface modification brought to the medical grade implants, by coating with the bionanocomposite, was also put under SEM observation. The structure and physical properties were studied using X-ray diffraction (XRD) (XRD, Philips,

Xpert, Cu Kα) at a scanning speed of 2°/min. Differential scanning calorimetry (DSC) assessment was done using (Mettler Toledo DSC 25), while to investigate the thermal and other physical properties, thermogravimetric/differential thermal analyses (TG/DTA) of the bionanocomposite material were carried out using (Perkin Elmer, Pyris Diamond, Tg-DTA high temp 115V) TGA instrument.

4.2.6 SWELLING PARAMETERS

A piece of approximately 1 cm^2 of the bionanocomposite film, weighing 0.0151 g, was allowed to swell in phosphate buffer saline solution (PBS) at room temperature. After immersion in PBS, the bionanocomposite film was removed at different time intervals and blotted with filter paper for the removal of excess PBS on the surface of the film. The weights of the dry and swollen films were measured using an electronic balance. Another approximately 1 cm^2 piece of the bionanocomposite film, weighing 0.0111 g, was also equilibrated in deionized water at room temperature, and weights of swollen and dry bionanocomposite film were also determined in the above mentioned manner. The swelling ratio (SR) of the Ag/CS bionanocomposite, at experimental temperature and different time intervals, was calculated as:

$$\text{Swelling ratio } (SR) = \left[(W_t - W_d)/W_d \right];$$

where W_t denotes weight of swollen material at time (t); and W_d denotes weight of dry material (before swelling).

4.2.7 QUANTITATIVE EVALUATION OF AG$^+$ RELEASE

Atomic absorption spectroscopy (AAS) was used for the quantitative determination of the silver ion concentration in the analyte. The analyte was prepared by taking 250 mL of phosphate buffer saline solution (PBS) pH ~ 7.0, which closely resembles human extracellular body fluid; in this buffer, Ag/CS bionanocomposite film weighing 10 mg was dipped for 48 h. Before dipping in the buffer the bionanocomposite film was washed several times with deionized water so as to remove any Ag$^+$ ion sticking to film. The analyte in the volume of 100 mL was analyzed through AAS using the spectrophotometer (ELICO Ltd. SL 194) for the estimation of Ag$^+$ released by the bionanocomposite film.

4.2.8 MECHANICAL TESTING

Mechanical properties of the material, bionanocomposite film was measured using (INSTRON 1195, Universal Testing Machine UTM, Buckinghamshire, England) running at a crosshead speed of 0.5 mm/min. The sample material was cut into 21 × 17 mm size and the mechanical testing was done using 0.100 kN load cell for tensile parameters; maximum stress and % elongation at break, which were measured and plotted.

4.2.9 SCRATCH TEST

Qualitative adhesion test was carried out to measure the adhesive strength of the Ag/CS bionanocomposite coating onto the substrate. This test measures the force needed to peel off the coating from the substrate under a contact load [27]. An especially designed scratch test apparatus was fabricated suitably for the purpose of measuring the scratching force. The scratch tool was made of stainless steel and two strain gages were used.

4.2.10 ANTIMICROBIAL ASSAY

The bionanocomposite was assayed for antimicrobial activity against *P. aeruginosa* (Gram negative) and *S. aureus* (Gram positive) microbial strains. Disc diffusion method [28] was used to find out the standard zone of inhibition (ZOI). Antibacterial test was done against the Ag/CS bionanocomposite film and medical grade stainless steel rod (30 × 12 mm) coated with the bionanocomposite. The film was cut into disc shape having 5-mm diameter, sterilized by UV radiation for 30 min and placed on cultured agar plate. Foley's catheter (disc-shaped pieces 6 mm in diameter) as a biomaterial, 2% CS film and plant extract of *Pseudotsuga menziesii* were observed as controls. Muller Hinton agar was used as culture media and inoculated with 300 µL of bacterial organism containing broth. The plates containing the bacterial and test samples were incubated at 37°C for 48 h. The plates were then examined for evidence of ZOI, which appear as a clear area around the discs. A meter ruler was used to measure the diameter of such ZOI.

4.2.11 EFFICACY OF THE AG/CS COATING ON STAINLESS STEEL SURFACE AGAINST BIOFILM FORMATION

4.2.11.1 Development of Biofilm

First, *S. aureus* culture, which resembles biofilm conditions of the bacterial strain, was developed in a 250-mL conical flask having mouth sealed with a cotton plug. For this purpose, *S. aureus* was inoculated into 50-mL sterile nutrient broth culture media. The culture was grown at 37°C, in a rotary shaker incubator, for 24 h. To assess the susceptibility of biofilm formation on the material surface, medical grade stainless steel implant plate piece (2.0 × 1.2 cm), coated with the Ag/CS bionanocomposite material, marked as (S), was suspended completely into the 50-mL *S. aureus* culture. The control consists of an uncoated piece of medical grade stainless steel plate (SS); of the same size, and also exposed to this particular *S. aureus* culture conditions. The entire set-up was again allowed to incubate at 37°C, in a rotary shaker incubator (60 rpm), for 24 h. The development of culture and biofilm environment conditions was initially indicated by the appearance of turbidity in the culture media, and later confirmed with the help of colony counter, showing the average cell density of ~7.13 $\log_{10}$ cfu cm^{-2}; the concentration of the bacterial inoculum was ~10^9 cfu mL^{-1}. Biofilm is a sessile microbial community embedded in a self-produced extracellular matrix.

4.2.11.2   Preparation of Samples for Study through SEM Observations

The coated and uncoated medical grade implant pieces (S) and (SS), respectively, were taken out of the suspended *S. aureus* culture and washed properly with phosphate buffer saline (PBS; pH~7.0) solution to remove the unbound bacteria. Then the sample pieces were dipped in 6.0% glutaraldehyde solution for 15 min in order to allow prefixing and cross-linking of the bound micro-organisms. The excess of glutaraldehyde was removed by thorough rinsing with PBS solution. There after dehydration of the samples was done through rinsing in 60%, 70%, 80%, 90%, 95% and 100% acetone. After this, the samples were taken for studying the efficacy against the biofilm formation, under SEM (JEOL JXA 8100).

4.2.12   IN VITRO BLOOD COMPATIBILITY TEST

Four tubes were taken and marked as '1, 2, 3 and 4,' containing 1 mL each of anti-coagulated human blood. The test sample, a piece of 5 mm × 5 mm Ag/CS bionanocomposite film weighing 0.0007 g, was then exposed to the anticoagulated human blood by suspending the sample material in tube (1), and 5 mm × 5 mm sample piece of only CS film weighing 0.0010 g was suspended in tube (2), with the help of sterilized threads. For control in tube (3), a single strand of sterilized thread alone was suspended, whereas for reference tube (4) was placed without any material and thread while contained with only 1 mL of anti-coagulated human blood. After observation and exposure of the samples for 1 h, they were taken out.

Blood cell counts (leukocytes and erythrocytes), platelets and hemoglobin were measured using the Beckman Coulter Counter (HMX) hematology analyzer. The degree of hemolysis was estimated by spectrophotometry through optical density measurement of blood plasma against normal saline, at wavelengths scanning from 350 to 650 nm, with the help of the instrument (Erba CHEM-5 Plus v2). The level of hemolysis was determined based on the absorbance peak of free hemoglobin at 414 nm.

4.2.13   CYTOTOXICITY TESTING

In vitro cytotoxicity test was performed through MTT assay against the J774A.1 murine macrophage cell line, as a parameter for the determination of the extent of biocompatibility of the developed Ag/CS bionanocomposite. The nanomaterial was placed as the test sample, whereas for the positive control macrophage cells only and for the negative control RPMI media only were considered.

4.3   RESULTS AND DISCUSSION

The mechanism of the formation of Ag/CS bionanocomposite can be proposed through the following reactions and schematic representations given in Figure 4.1.

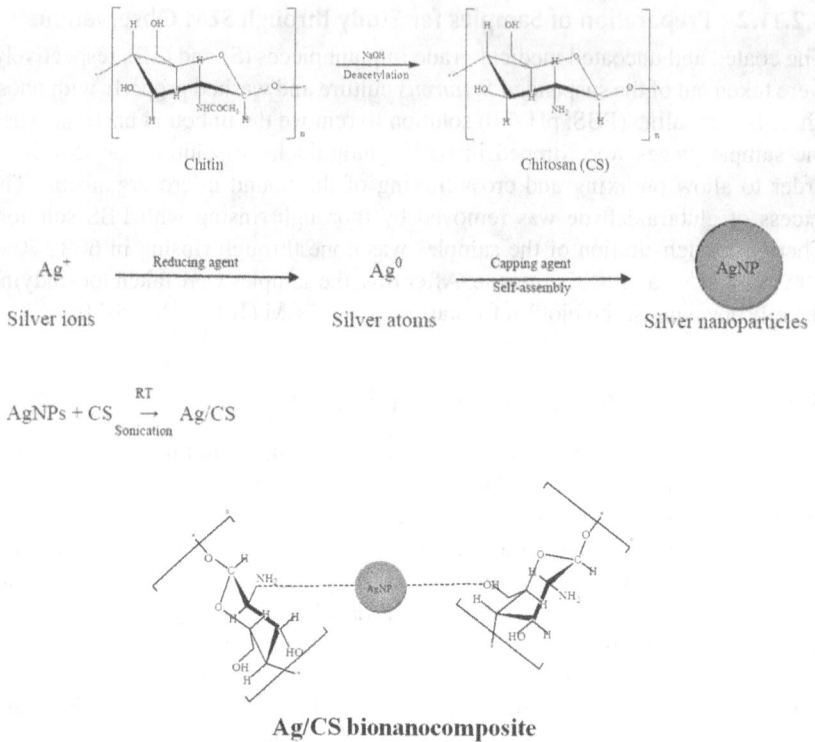

Ag/CS bionanocomposite

FIGURE 4.1 Reaction mechanism for the formation of Ag/CS bionanocomposite.

4.3.1 FTIR ELUCIDATION

4.3.1.1 FTIR Study of the Silver Nanoparticles

In Figure 4.2, the FTIR spectrum of silver nanoparticles (AgNPs R5) elucidates the active chemical species in the foliage needle extract that are involved in the phytofabrication of silver nanoparticles. The clearly observed large –OH and –CH stretches (3417 and 2919 cm^{-1}) are characteristic of sugar molecules adhered to the silver nanoparticles. Primary nutrients (soluble carbohydrates), especially sugars like glucose, fructose and galactose present in the needle tissue of the plant *Pseudotsuga menziesii* [29], are readily extracted in the aqueous solution and likely function as both reducing and stabilizing moieties [30, 31]. The peak at (1619 cm^{-1}) can be attributed predominantly to the overlapping stretching vibrations of –C=C– and –C=O characters probably derived from aromatic rings or polyphenolics, e.g. flavonoids, or heterocyclic compounds, e.g. vitamin C [32]. The peak at (1384 cm^{-1}) may be ascribed to –C–N stretching modes of the amine [33]. The strong peak present at (1077 cm^{-1}) corresponds to the stretching vibrations of –C–O–C group or –C–O carboxyl group of amino acids, whereas the broadband at (616 cm^{-1}) is due to –N–H bending vibrations. FTIR studies confirm

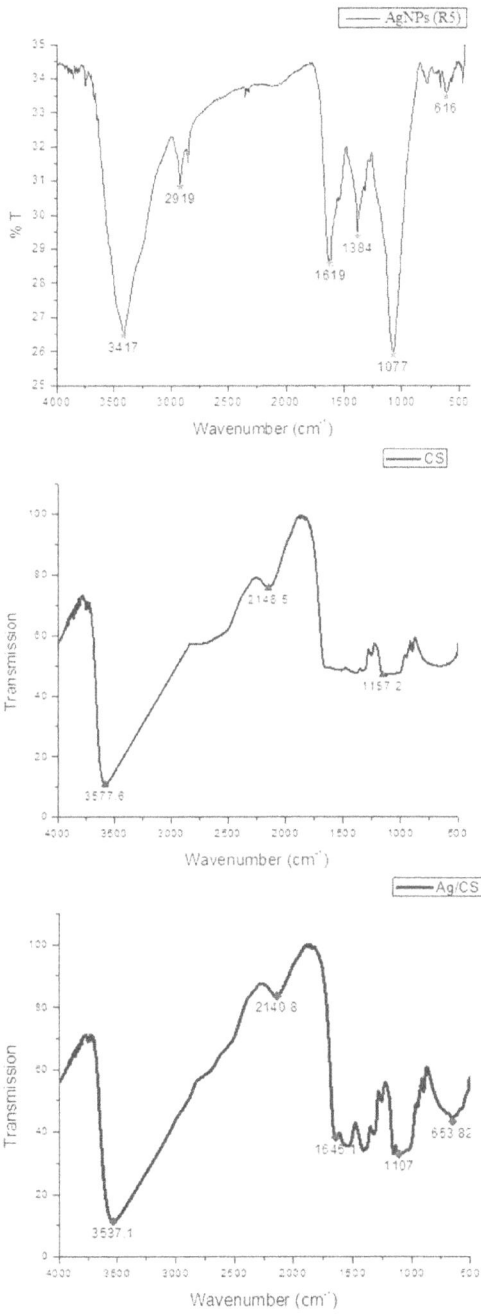

FIGURE 4.2 FTIR spectra (%Transmittance/Wavenumber cm⁻¹) of silver nanoparticles generated by *Pseudotsuga menziesii* foliage phytomass, CS film and Ag/CS bionanocomposite film, respectively.

that amino acids, protein residues, carbohydrates together with various plant secondary metabolites may also have the ability to act as capping ligands that are adsorbed on the surface of silver nanoparticles.

4.3.1.2 FTIR Study of the Ag/CS Bionanocomposite

In the FTIR spectrum of the Ag/CS bionanocomposite in Figure 4.2, the peak at ~3,500 is more pronounced corresponding to the axial OH group of the CS molecule. The bending vibrations between 1600 and 1000 cm^{-1} also intensify, indicating possible interaction between silver nanoparticles and amino group of CS [34–36].

4.3.2 SEM INVESTIGATION OF THE AG/CS NANOCOMPOSITE FOR MORPHOLOGICAL PROPERTIES

The Ag/CS bionanocomposite film was studied for morphological characteristics through SEM. In Figure 4.3, SEM micrograph shows nanoparticles well dispersed in CS matrix with minimum aggregation. The interaction between the lone pair of electrons at the amine group as well as on the hydroxyl group of CS and the partial positive charge developed at the surface of the silver nanoparticles due to electron drift [37] effectively stabilizes the silver nanoparticles and prevents agglomeration. The bionanocomposite suspension was able to be coated uniformly over the surface of medical grade implants used inside the human body, together with imparting smooth surface modification to the biomaterials, vividly elucidated through the SEM observations.

4.3.3 CHARACTERIZATION OF THE NANOCOMPOSITE THROUGH DSC, TG/DTA AND XRD

The enthalpy change of the Ag/CS bionanocomposite film with respect to temperature and time was investigated through DSC. The Ag/CS bionanocomposite film melts at ~106°C undergoing endothermic decomposition. The TG/DTA curve in Figure 4.3 illustrates the thermal properties of the film. There appears complexity in the thermal degradation process as revealed by the thermograms, resulting into unclear curves.

The DTA curve expresses the endothermic effect and denotes that till ~250°C mostly evaporation of the trace solvents (e.g. water and acetic acid) used for the preparation of Ag/CS bionanocomposite film takes place. The entire complex process of thermal decomposition is attributed to several conditions, initiating from dehydration of the saccharide rings, to decomposition and depolymerization of both acetylated as well as deacetylated units of CS. The decomposition of the bionanocomposite material occurs at stages and the process is probably thermo-oxidative. While the TG curve represents a very small slope, due to which no peak could be possibly identified suggesting any noticeable degradation stage. Still, it may be concluded that dispersing silver nanoparticles in CS matrix imparts reasonable thermal stability in comparison to only CS film according to the initial temperature at which degradation begins [23].

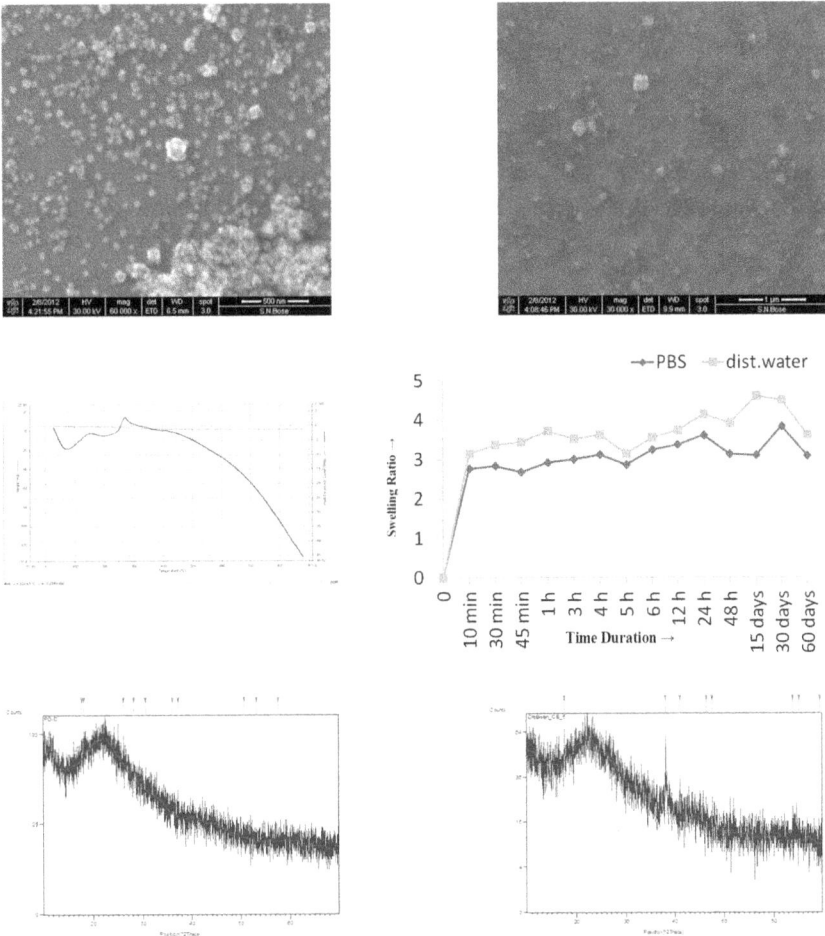

FIGURE 4.3 Morphological and physical characteristics of the developed Ag/CS bionanocomposite.

Figure 4.3 shows XRD pattern of pure CS film and Ag/CS bionanocomposite film. The diffractogram of the pure CS film indicates an amorphous nature, as reported in earlier studies [23], does not show any sharp or highly intense peaks but exhibits broad peaks at 2θ of $11°$ and $21°$. While in the diffraction pattern of Ag/CS bionanocomposite film there is development of crystallinity in the amorphous polymer indicating a semi-crystalline structure. Three diffraction peaks observed at $2\theta = 38°$, $43°$ and $64°$ correspond to the (111), (200) and (220) Miller indices crystallographic planes, respectively, of the face-centered cubic (FCC) Ag°, (JCPDS file No. 00-004-0783). The peaks obtained are not very sharp, due to the capped silver nanoparticles by the plant biomolecules on the surface. The two broad peaks below 30° also exist, which indicates the presence of the amorphous CS component.

4.3.4   SWELLING PARAMETERS

The swelling capacity of a bionanocomposite material plays an important role in the antibacterial activity as well as in the wound healing capacity due to its high water/solvent holding capacity. It can further absorb moderate amount of the wound exudates by swelling which helps in fast healing. The developed Ag/CS bionanocomposite material also contains free $-NH_2$ groups which gain positive charge when dissolved in acidic medium. Swelling of the bionanocomposite film is chiefly influenced by interactions between CS chains. *SR* of the bionanocomposite film, though having slight erratic pattern, confirmed the satisfactory solvent uptake like water or other biological fluids, without dissolving in them. The swelling behavior with time is clearly shown through the graph in Figure 4.3.

4.3.5   QUANTITATIVE ESTIMATION OF AG+ RELEASE

Ag^+ present in 100 mL of the analyte solution taken for the test was < 0.1 ppm, which was the lower detection limit of the instrument. It was observed through antimicrobial studies that the Ag/CS bionanocomposite film release Ag^+ in a concentration level capable of rendering antimicrobial efficacy, and in amounts in which it is nontoxic to the human cells and therefore safe.

4.3.6   MECHANICAL PROPERTIES

The mechanical properties achieved by the developed Ag/CS bionanocomposite film have been presented in Figure 4.4. It has been reported earlier that reinforcement of CS films imparts better mechanical properties on several parameters [38, 39]. This fact is supported by the present data showing positive effect of the Ag/CS on the tensile strength. The phenomenon suggests that incorporating nano-sized fillers into CS matrix promotes rigidity, as a consequence of intense interactions between the reinforced particles and the matrix, but may also result into reduction of elongation at break [40].

4.3.7   ASSESSMENT OF THE AG/CS NANOCOMPOSITE
COATED SURFACE THROUGH SCRATCH TEST

The experiment was done in Mecmesin using 500 N Load Cell with two special grips to hold the substrate in position. A special stainless steel surgical scraper tool was applied to scratch the surface. Applying the scratch test apparatus for the determination purpose, it was found that a sufficient force is needed to peel off the Ag/CS bionanocomposite coating from the substrate. The thickness of the bionanocomposite coating measured is 25 ± 0.1 μm. On an average 6 N force was required to scratch a width of 1.9-mm and 25×10^{-3}-mm thickness. Therefore,

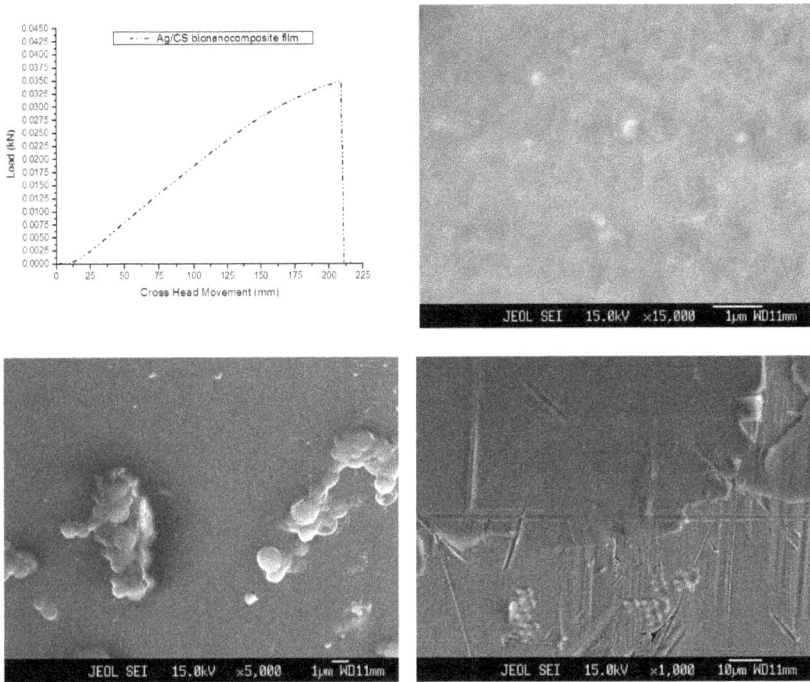

FIGURE 4.4   Illustration of the mechanical property and bioactivity of the developed biocompatible and bioactive Ag/CS bionanocomposite.

mean coating strength was measured to be 126.3 MPa. The scratch strength was measured by applied stress, equal to F/A.

$$F = 6N \quad A = 1.9 \times 0.025 \, mm^2$$

$$\text{Strength } F/A = 6/0.0475 = 126.3 \text{ MPa}$$

4.3.8   ANTIMICROBIAL ASSAY

Ag/CS bionanocomposite film and stainless steel plate coated with the bionanocomposite were assayed for antimicrobial activity against *P. aeruginosa* (Gram nega-tive) and *S. aureus* (Gram positive) microbial strains, which cause majority of the biomedical implant-related infection. Disc diffusion protocol was adopted to deter-mine the standard ZOI. Details of the result obtained are listed in Table 4.1. It has been observed in the present study that the effect was well pronounced against Gram-negative bacteria which contain only a thin peptidoglycan layer of 2~3 nm between the cytoplasmic membrane and the outer membrane, and also against Gram-positive bacteria which lack the outer membrane but have a peptidoglycan layer of about 30-nm thickness.

TABLE 4.1

Zone of Inhibition (ZOI) in (mm) against Selective Bacterial Strains

Name of the Micro-organism	Ag/CS Film (1′)	Stainless Steel Plate Coated with Ag/CS(S)	Chitosan Film (CS)	Uncoated Biomaterial (UB)	Plant Extract (PE5)
Pseudomonas aeruginosa	18	NA[a]	Nil[b]	Nil[b]	Nil[b]
Staphylococcus aureus	25	33	NA[a]	NA[a]	NA[a]

Notes:

[a] NA stands not applicable (represents test not done).

[b] Nil represents no observed zone of inhibition (ZOI).

The antimicrobial activity of silver has been known since ages, when silver vessels were used for storing to prevent spoilage of food. Hippocrates recognized the role of silver in the prevention of disease and clinicians have accepted it for over 100 years or more. The mode of action has been studied recently in the last few decades, with the advent of nanotechnology and silver nanoparticles. Metallic silver when exposed to aqueous environment releases silver ions (Ag^+) which binds with the thiol groups of certain amino acids (as per the complex stability reports the heavy metal ions have more affinity for sulfur-containing ligands) and inhibits the enzymes of respiratory cycle and it also interferes with the DNA replication of the micro-organism. Smaller sized particles have higher surface to volume ratio and are thus more effective. The main mechanism exerting which silver nanoparticles manifested antibacterial activities was by anchoring to the bacterial cell wall and then penetrating into, modulating cellular signaling thereafter through dephosphorylating putative key peptide substrates on tyrosine residues. Finally causes cell lysis by rupturing the cytoplasmic membrane [41].

4.3.9 EFFICACY OF THE Ag/CS NANOCOMPOSITE COATING OVER STAINLESS STEEL AGAINST BIOFILM FORMATION

SEM observations revealed that the coated (S) and uncoated (SS), stainless steel medical grade implant pieces, suspended simultaneously in the *S. aureus* culture, resembling biofilm environmental conditions, had differential growth of biofilm on their surfaces. Coating of the Ag/CS bionanocomposite on implant piece (S) imparted smooth surface modification together with providing a self-sterilizing property, which drastically hindered the possibility of bacterial adhesion and therefore no successive biofilm formation on its surface. The antimicrobial coating restricted bacterial adhesion at the first place and smooth surface modification did not favor their proliferation at any case [8]. Whereas, the uncoated implant piece (SS) lacking these characteristics was much enough prone and susceptible

TABLE 4.2
Optical Density (OD) Reading

Tube	Sample	OD
1	Ag/CS film	0.099
2	CS film	0.111
3	Reference	0.136
4	Control	0.139

to the growth of the pathogenic microorganism; as it provided a favorable surface for bacterial adhesion and promoted their proliferation, therefore, showed substantial formation of biofilm on its surface. SEM micrographs displayed in Figure 4.4 depicts a part of the study to explain.

4.3.10 IN VITRO BLOOD COMPATIBILITY ASSAY

Ag/CS bionanocomposite material did not cause any noticeable hemolysis of the blood sample to which it was exposed, because a sample is classified as being hemolyzed if the A_{414} reading exceeds a value of 0.2 [42]. The readings of the performed optical density assessment are summarized in Table 4.2. When the cell count changes (leukocytes, erythrocytes and platelets), in reference is compared with the test sample, none of the parameters showed significant variation as an effect of material contact with blood. Results of the analysis are listed in Table 4.3. It is conclusively proved that the bionanocomposite is possibly possessed with factors reducing hemolysis while having good blood compatibility [43].

4.3.11 IN VITRO CYTOTOXICITY TEST

The synthesized Ag/CS bionanocomposite material can be considered reasonably biocompatible and safe to be used in regulated quantity on the basis of the cytotoxicity test performed. The CC_{50} value of the silver nanoparticles, obtained via phytofabrication from the plant *Pseudotsuga menziesii,* was

TABLE 4.3
Blood Cell Counts of the Samples

Tube	Sample	Leukocytes (WBC)	Erythrocytes (RBC)	Platelets (PLT)	Hemoglobin (HGB)
1	Ag/CS film	$6.10 \times 10^3/\mu L$	$4.23 \times 10^6/\mu L$	$168 \times 10^3/\mu L$	13.50 g/dL
2	CS film	$6.41 \times 10^3/\mu L$	$4.32 \times 10^6/\mu L$	$173 \times 10^3/\mu L$	13.61 g/dL
3	Reference	$6.20 \times 10^3/\mu L$	$4.12 \times 10^6/\mu L$	$162 \times 10^3/\mu L$	13.20 g/dL
4	Control	$6.40 \times 10^3/\mu L$	$4.06 \times 10^6/\mu L$	$161 \times 10^3/\mu L$	12.90 g/dL

0.282304 ± 0.000644 µg/mL and of the Ag/CS bionanocomposite material was 0.040 ± 0.006 mg/mL while that of the only CS matrix was 2.892 ± 0.199 mg/mL.

4.4   PRINCIPLE OF APPLICATION

Through this study we developed a vivid conception that coating of medical implants with the self-sterilizing Ag/CS bionanocomposite nanobiomaterial can definitely restrict bacterial adhesion and their subsequent biofilm formation, with high efficacy, strongly minimizing the risks of the stagnant and increasingly haunting problem of implant-related infections [44, 45]. We have hauled here herbal route to synthesize silver nanoparticles for the purpose, in order to add the beneficiaries of herbs, while discarding toxic and harmful reagents which are obnoxious to be handled as well as to be applied, as they persistently adhere to the surface of the nanostructures. This is the chief advantage of this herbal route over common methodologies and other synthetic route [46], which may offer better monodispersity control. This herbal method strictly sticks to the principles drafted for green synthesis [47, 48] despite a wide range in shape and size of the nanoparticles may be achieved.

The free flowing Ag/CS bionanocomposite material finally formed can be conveniently used for antimicrobial coating over biomaterials [49–51]. This nanobiomaterial is also biocompatible and biodegradable, therefore, will cause minimal harm to the human body even after being leached out [52]. While, at the same time it imparts smooth, self-sterilizing surface to the implant biomaterials at the time of being incorporated into the body, to combat microorganisms at that particular instant and control the susceptibility to BAI. The developed nanocomposite surface coating system from our research may also be drawn to wider applications for surrounding object surfaces having metallic, ceramic, polymeric, wooden, etc. texture, such as of railings, switches, touch-screens, table-tops, door handles, etc., in order to minimize microbial contamination through common handling, thus restricting community infections caused by viruses, bacteria and fungi [53–55].

4.5   CONCLUSION

Thus, this first attempt to biosynthesize Ag/CS bionanocomposite through herbal route via the intervention of the plant *Pseudotsuga menziesii* shows significant potentiality to mitigate the menace of medical implant associated infections. Taking into account, there is plenty of room in the garden and at the bottom; this biodegradable, bactericidal, self sterilizing, biocompatible material holds good potential to ameliorate the face of existing biomaterials through surface modification in a cost effective way.

REFERENCES

1. G. G. Geesey, Cur. Opin. In Microbiol., 4 (2001) 296–300.
2. P. Lejeune, Trend in Microbiol., 11 (2003) 179–184.

3. N. Hoiby, T. Bjarnsholt, M. Givskov, S. Molin and O. Ciofu, Int. J. Antimicrob. Agents, 35 (2010) 322–332.
4. C. Von Eiff, B. Jansen, W. Kohnen and K. Becker, Drugs, 65 (2005) 179–214.
5. C. Von Eiff, G. Peters and C. Heilmann, Lancet. Infect. Dis., 2 (2002) 677–685.
6. J. L. Vincent, Lancet, 361 (2003) 2068–2077.
7. G. M. Dickinson and A. L. Bisno, Antimicrob. Agents Chemother., 33 (1989) 602–607.
8. M. Katsikogianni and Y. F. Missirlis, Eur. Cells Mater., 8 (2004) 37–57.
9. B. Gottenbos, H. J. Busscher, H. C. Van der Mei and P. Nieuwenhuis, J. Mat. Sci.: Mat. In Med., 13 (2002) 717–722.
10. F. Gotz, Mol. Microb., 43 (2002) 1367–1378.
11. P. E. Vaudaux, F. A. Waldvogel, J. J. Morgenthaler and U. E. Nydegger, Infect. Immunol., 45 (1984) 768–774.
12. H. H. M. Rijnaarts, W. Norbe, E. J. Bouwer, J. Lyklema and A. J. B. Zehnder, Col. Surf. B, 4 (1995) 5–22.
13. J. Cordero, L. Munuera and M. D. Folgueira, Injury, 27 (1996) 34–37.
14. T. R. Scheuerman, A. K. Camper and M. A. Hamilton, J. Col. Interf. Sci., 208 (1998) 23–33.
15. V. A. Tegoulia and S. L. Cooper, Col. Surf. B, 24 (2002) 217–228.
16. D. J. Balazs, K. Triandafillu, Y. Chevolot, B. O. Aronsson, H. Harms, P. Descouts and H. J. Mathieu, Surf. Interf. Anal., 35 (2003) 301–309.
17. G. Speranza, G. Gottardi, C. Pederzolli, L. Lunelli, R. Canteri, I. Pasquardini, E. Carli, A. Lui, D. Mniglio, M. Brugnara and M. Anderle, Biomaterials, 25 (2004) 2029–2037.
18. E. A. Deitch, A. A. Marino, V. Malakanov and J. A. Albright, J. Trauma, 27 (1987) 301–304.
19. S. Saint, J. G. Elmore, S. D. Sullivan, S.S. Emerson and T. D. Koepsell, Am. J. Med., 105 (1998) 236–241.
20. R. M. Joyce-Wohrmann, T. Hentschel and H. Munstedt, Adv. Eng. Mater., 2 (2000) 380–386.
21. U. Klueh, V. Wagner, S. Kelly, A. Johnson and J. D. Bryers, J. Biomed. Mater. Res., 53 (2000) 621–631.
22. T. Kocourek, M. Jelínek, J. Mikšovský, K. Jurek and M. Weiserová, J. Phys., 497 (2014) 1–9.
23. L. Balau, G. Lisa, M. I. Popa, V. Tura and V. Melnig, Cent. Eur. J. Chem., 2 (2004) 638–647.
24. Z. Chengjun and W. Qinglin, Colloids Surf. B, 84 (2011) 155–162.
25. S. K. Mazumder, Composites Manufacturing, Materials, Product and Process Engineering, CRC Taylor & Francis, London (2002) ISBN 0-8493-0585-3.
26. J. W. Rhim, C. L. Weller and K. S. Ham, Food Sci. Biotechnol., 7 (1998) 263–268.
27. S. Pal, S. Roy and S. Bag, Trends Biomater. Artif. Organs, 18 (2005) 106–109.
28. R. Cruickshank, Medical Microbiology: A Guide to Diagnosis and Control of Infection (11th ed.), E. & S. Livingston Ltd, Edinburgh and London (1968) 888.
29. J. Zou and R. G. Cates, J. Chem. Ecol., 21 (1995) 387–402.
30. J. Liu, M. Anand and C. B. Roberts, Langmuir, 22 (2006) 3964–3971.
31. S. K. Mehta, S. Chaudhary and M. Gradzielski, J. Col. Inter. Sci., 343 (2010) 447–453.
32. C. Ramteke, T. Chakrabarti, B. K. Sarangi and R. A. Pandey, J. Chem., Article ID 278925 (2013) 1–7.
33. S. M. Ghoreishi, M. Behpour and M. Khayatkashani, Physica E., 44 (2011) 97–104.

34. S. Govindan, E. A. K. Nivethaa, R. Saravanan, V. Narayanan and A. Stephen, Appl. Nanosci. (2012) 1–5.

35. P. Kaur, A. Choudhary and R. Thakur, Int. J. Sci. Eng. Res., 4 (2013) 869–872.

36. S. Akmaz, E. Dilaver, A. Jgüzel, M. Yasar and O. Erguven, Adv. Mat. Sci. Eng., Article ID 690918 (2013) 1–6.

37. S. W. Kang, K. Char and Y. S. Kang, Chem. Mater., 20 (2008) 1308–1311.

38. K. A. M. Amin and M. in het Panhuis, Polymers, 4 (2012) 590–599.

39. P. Cazón and M. Vázquez, Environ. Chem. Lett., 18 (2020) 257–267.

40. S. Azizi, M. B. Ahmad, M. Z. Hussein, N. A. Ibrahim and F. Namvar, Int. J. Nanomed., 9 (2014) 1909–1917.

41. S. Shrivastava, T. Bera, A. Roy, G. Singh, P. Ramchandrarao and D. Dash, Nanotechnology, 18 (2007) 225103–225111.

42. M. B. Kirschner, J. J. B. Edelman, S. C. H. Kao, M. P. Vallely, N. V. Zandwijk and G. Reid, Front. Genet., 4 (2013) 94–101.

43. L. A. Harker, B. D. Ratner and P. Didisheim, Cardiovasc. Pathol., 2(Suppl 3)(1993) 1S–224S.

44. S. Samuggam, S. V. Chinni, P. Mutusamy, S. C. B. Gopinath, P. Anbu, V. Venugopal, L. V. Reddy and B. Enugutti, Molecules, 26 (2021) 2681, 1–13.

45. S. Kligman, Z. Ren, C. H. Chung, M. A. Perillo, Y. C. Chang, H. Koo, Z. Zheng and C. Li, J. Clin. Med., 10 (2021) 1641, 1–36.

46. A. Panja, A. K. Mishra, M. Dash, N. K. Pandey, S. K. Singh and B. Kumar, EJMO, 5 (2021) 95–102.

47. R. Perveen, S. Shujaat, M. Naz, M. Z. Qureshi, S. Nawaz, K. Shahzad and M. Ikram, Mater. Res. Express, 8 (2021) 055007, 1–11.

48. I. Ahamad, N. Aziz, A. Zaki and T. Fatma, J. Appl. Phycol., 33 (2021) 829–841.

49. P. Dwivedi, S. S. Narvi and R. P. Tewari, IJBNN, 2 (2012) 187–206.

50. P. Dwivedi, S. S. Narvi and R. P. Tewari, Int. J. Eng. Res. Appl., 2 (2012) 1490–1495.

51. P. Dwivedi, S. S. Narvi and R. P. Tewari, Nano LIFE, 5 (2015) 1540006.

52. D. Bamal, A. Singh, G. Chaudhary, M. Kumar, M. Singh, N. Rani, P. Mundlia and A. R. Sehrawat, Nanomaterials, 11 (2021) 2086, 1–40.

53. R. Z. Adam and S. B. Khan, PLoS ONE, 16 (2021) e0245811, 1–13.

54. N. Jain, P. Jain, D. Rajput and U. K. Patil, Micro and Nano. Syst. Lett., 9 (2021), 1–24.

55. P. Dwivedi, D. Tiwary, S. S. Narvi, R. P. Tewari and K. P. Shukla, Lett. Appl. Nanobiosci., 9 (2020) 1485–1493.

5 Bioactive Silver/ Chitosan-g-Polyacrylamide (Ag/CS-g-PAAm) Nanocomposite Hydrogel as Super Absorbent Polymeric (SAP) Material

Phytomass Intervened Value Conversion

5.1 INTRODUCTION

Biomaterials associated infections (BAI), which occur in approximately 0.5–6% [1, 2] of all cases, strongly depending on the implant site, are difficult to treat, as the bacterial biofilm mode of growth protects the infecting organisms against the host immune system and antibiotic treatment [3]. Today the advent of multi-drug resistant 'super bugs', like the one recently discovered, 'NDM-l' New Delhi Metallo-beta-lactamase, have made the problem of BAI much more pronounced [4]. Efforts and researches in this direction to minimize infection by designing self-sterilizing materials are in progress. The most common method to synthesize 'antimicrobial materials' is the addition of an active agent, such as silver [5], quaternary ammonium salts, phenols and antibiotics to a polymer as an additive [6, 7]. These additives, which are biocides and are capable of killing all living organisms, are slowly leached out in the surrounding biological environment.

Silver and silver ion-based materials are well known for their antibacterial and antifungal activity [8, 9], as well as its use as prostheses, catheters, vascular grafts and as wound dressings [10, 11]. Researchers have reported that silver in the

DOI: 10.1201/9781003217343-5

nanostructure form is increasingly effective as antimicrobial agent, in comparison to the bulk counterpart. The therapeutic efficacy of silver in nanoparticles form is several folds higher than conventional silver compounds [12]. In addition, silver nanoparticles-based antimicrobial materials have further advantages due to their thermal stability, health and relative environmental safety. Many workers have reported methods to synthesize composites of silver nanoparticles and polymers, which act as antimicrobial materials. Based on this, various natural and synthetic polymers, like starch, gelatine, sodium alginate, carboxy methyl cellulose, chitosan (CS), poly(vinyl alcohol), poly(vinyl chloride), poly(methyl methacrylate), poly(acryl amide), etc., have been employed to prepare biocompatible polymeric silver nanocomposites [13].

The main aim over here is to develop material for biomedical applications which will show self-sterilizing antibacterial properties, but when leached out will produce minimal harm to the human body. Therefore, a novel silver nanocomposite has been synthesized, having silver nanoparticles dispersed in chitosan-g-polyacrylamide hydrogel matrix (Ag/CS-g-PAAm), through herbal method for the purpose. This is a novel experimental study employing herbal route for the nanobiotechnological synthesis and phytomass intervened value conversion into this silver nanocomposite material and applying it as a biomaterial for biomedical engineering, having quick prospects for combating BAI.

Over here for the fabrication of silver nanoparticles, fresh rhizome of the perennial herb *Curcuma longa* was used which contains curcumin as its active ingredient, having anti-inflammatory, anti-oxidative, anticancerous and anti: -viral, -fungal, -bacterial effects [14, 15]. The hydrogel matrix having the biopolymer CS, composed of poly(β-(1-4)-2-amino-2-deoxy-D-glucose), is one of the structural polysaccharide which is abundantly available in nature after cellulose [16–18] will stabilize the metallic nanoparticles [19]. Silver nanoparticles will release silver ions in the presence of H_2O molecules, which are highly antimicrobial, oligodynamic and also nontoxic to human cells at lower concentrations.

The modification of CS with acrylamide (AAm), to (CS-g-PAAm) hydrogel [20], enhances the hydrophilicity of the matrix and mechanical strength of the nanocomposite biomaterial [21]. The developed Ag/CS-g-PAAm hydrogel was investigated through several characterization techniques together with antimicrobial analysis. This material is cost efficient and should be biocompatible, as it has been synthesized via nanobiotechnological phytofabrication route, eliminating conventional methods, which give rise to physiologically and environmentally hazardous by-products, intense heat, as well as draw potentially toxic chemicals.

5.2 EXPERIMENTAL SECTION

5.2.1 MATERIALS

CS (degree of deacetylation: 79%, molecular mass: 500,000 g/mol) was purchased from Sea Foods (Cochin), India; acrylamide (AAm), ammonium persulfate (APS), N,N,N',N'-tetramethylethylenediamine (TEMED) and

N,N'-methylenebisacrylamide (MBA) from Sisco Research Laboratory, India; acetic acid glacial (extra pure) and nutrient agar from Thomas Baker (Chemical) Pvt. Ltd., India. Silver nanoparticle suspension from reaction mixture (R9), phytofabricated by the rhizome of the herb *Curcuma longa* (Turmeric). Bacterial strains *Pseudomonas aeruginosa* (Gram negative), *Escherichia coli* (Gram negative) and *Staphylococcus aureus* (Gram positive) were obtained from the culture bank of Microbiology Department, Sam Higginbottom Institute of Agriculture, Technology and Sciences, Allahabad, India. All the solutions were prepared with the use of deionized water as solvent.

5.2.2 SYNTHESIS – SYNTHESIS OF AG/CS-G-PAAM NANOCOMPOSITE HYDROGEL SUPER ABSORBENT POLYMERIC (SAP) MATERIAL

The suspension of silver nanoparticles (RM9), phytofabricated by fresh rhizome of *Curcuma longa* (experimental details referred and discussed in Chapter 3), was used because of the number of medicinal benefits present in turmeric extract and a part of it assumed to be adhering to the nanoparticles mediated through it. After complete stabilization of nanoparticles was confirmed, the reaction mixture (RM9) was centrifuged at 9,000 rpm for 15 min and the residue of silver nanoparticles were re-dispersed in distilled water. This procedure was repeated three times to isolate and purify the nanoparticles from other unbound biomolecules. The obtained residue of the silver nanoparticles from RM9 denoted as (R9) (in this chapter) was dispersed in 30 mL of CS solution (2% [w/v] in 1% [v/v] acetic acid).

This dispersion mixture prepared was put inside an RB flask placed into a water bath which was preset at 60°. Into the mixture added 0.10 g of APS and stirred for around 10 min. Following this, 2.0 g of AAm was added and thereafter MBA 0.05 g was added as a crosslinker. The reaction mixture was continuously stirred for 1 h under nitrogen atmosphere. The resulting hydrogel was neutralized to pH 8 by the addition of 1N NaOH solution [20]. Then methanol was added to the gel product while stirring and after complete dewatering the transparent orange nanocomposite gel was dried at room temperature.

5.2.3 FTIR SPECTROSCOPY

5.2.3.1 FTIR Study of the Silver Nanoparticles

The dried residue of silver nanoparticles (R9), obtained from RM9, phytofabricated by rhizome extract of *Curcuma longa*, was characterized using Fourier transformed infrared spectroscopy (FTIR). The FTIR analysis was done to have an idea of the possible biomolecules responsible for the reduction of Ag^+ ions and identify the functional groups capping and efficiently stabilizing the silver nanoparticles. This also gives an account of the beneficial compounds of the plant adsorbed on the surface of the nanoparticles as an organic coat. The isolated and purified residue of silver nanoparticles from other unbound bio-organic compounds also present in the suspension were received after repeated centrifugation. FTIR spectrum of these silver nanoparticles was recorded over the range of (400–4000) cm^{-1} with (PerkinElmer) spectrophotometer for studying the chemical properties.

5.2.3.2   FTIR Study of the Ag/CS-g-PAAm Nanocomposite

To assess the chemical structure of the nanocomposite biomaterial, FTIR spectrum was also recorded with the spectrophotometer (FTLA 2000 ABB) over the range of (400–4000) cm^{-1}.

5.2.4   CHARACTERIZATION OF AG/CS-G-PAAM NANOCOMPOSITE

The synthesized Ag/CS-g-PAAm nanocomposite hydrogel was characterized using scanning electron microscopy (SEM), X-ray diffraction (XRD), differential scanning calorimetry (DSC) and thermo-gravimetric/differential thermal analysis (TG/DTA). DSC analysis was done (using Metler Toledo DSC 25) and TG/DTA of the biomaterial were carried out using (Perkin Elmer, Pyris Diamond, Tg-DTA high temp 115V) TGA instrument. The structure of the nanocomposite biomaterial was studied using XRD (XRD, Philips, Xpert, Cu Kα) at a scanning speed of 3°/min. SEM analysis was carried (through JEOL JXA 8100) to perform the morphological investigation of the nanobiomaterial.

5.2.5   MECHANICAL TESTING

Mechanical properties of Ag/CS-g-PAAm nanocomposite were measured using (INSTRON 1195, Universal Testing Machine UTM, Buckinghamshire, England) running at a crosshead speed of 0.5 mm/min. The sample material was cut into 20×50-mm size and the gauge length was about 10 mm. The mechanical testing was done using 20 kN load cell for tensile parameters, maximum stress and % compression at break, which were measured and plotted.

5.2.6   SWELLING PARAMETERS

The pre-weighed and dried Ag/CS-g-PAAm nanocomposite hydrogel samples, weighing 0.1196 and 0.0580 g, were equilibrated in 250 mL of phosphate buffer saline solution (PBS) pH ~7.0 and in 250 mL of deionized water respectively, at room temperature. After immersion in PBS and deionized water separately the nanocomposite samples were removed at different time intervals and blotted with filter paper for the removal of excess solvent from the surface. The solvent uptake by the samples at time intervals were measured using an analytical electronic balance. The swelling ratio (SR) of the nanobiomaterial was calculated applying the equation:

$$\text{Swelling ratio } (SR) = \left[(W_t - W_d) / W_d \right];$$

where W_t denotes weight of swollen material at time (t); and W_d denotes weight of dry material (before swelling).

5.2.7   QUANTITATIVE EVALUATION OF Ag^+ RELEASE

Atomic absorption spectroscopy (AAS) was used for the quantitative determination of the silver ion concentration in the analyte. The analyte was prepared by taking 250 mL of PBS pH ~7.0, which closely resembles human extracellular body fluid; in this Ag/CS-g-PAAm nanocomposite hydrogel weighing 24.1 mg was dipped for 48 h. Out of this analyte, 100 mL was analyzed through an atomic absorption spectrophotometer (ELICO Ltd. SL 194) for the estimation of Ag^+ released by the nanocomposite material.

5.2.8   ANTIMICROBIAL ASSAY

The nanocomposite having the bioreduced silver was assayed for antimicrobial activity against *P. aeruginosa* (Gram negative), *E. coli* (Gram negative) and *S. aureus* (Gram positive) microbial strains. Disc diffusion protocol was used to measure the standard zone of inhibition (ZOI). The plates containing the bacterial and test samples were incubated at 37°C for 48 h. The plates were then examined for evidence of ZOI and measured using a meter ruler.

5.2.9   *IN VITRO* BLOOD COMPATIBILITY TEST

Three tubes were taken and marked as '1, 2, 3', containing 1 mL each of anti-coagulated human blood. The test sample, a piece of Ag/CS-g-PAAm nanocomposite material weighing 0.0246 g, was then exposed to the anti-coagulated human blood by suspending the sample material in tube (1) with the help of a sterilized thread. A strand of sterilized thread alone was suspended for control in tube (2), whereas tube (3) was used for reference without any material and thread while containing only 1 mL of anti-coagulated human blood. After observation and exposure of the samples for 1 h, they were taken out.

Blood cell counts (leukocytes, erythrocytes and platelets) were measured using Beckman Coulter Counter (HMX) hematology analyzer. The degree of hemolysis was estimated by spectrophotometry through optical density measurement of blood plasma against normal saline, at wavelengths scanning from 350 to 650 nm, with the help of the instrument (Erba CHEM-5 Plus v2). The level of hemolysis was determined based on the absorbance peak of free hemoglobin at 414 nm.

5.3   RESULTS AND DISCUSSION

The general mechanism of graft copolymerization process initiated by APS for the synthesis of Ag/CS-g-PAAm nanocomposite hydrogel in the presence of MBA can be represented through the following reactions and scheme given in Figure 5.1.

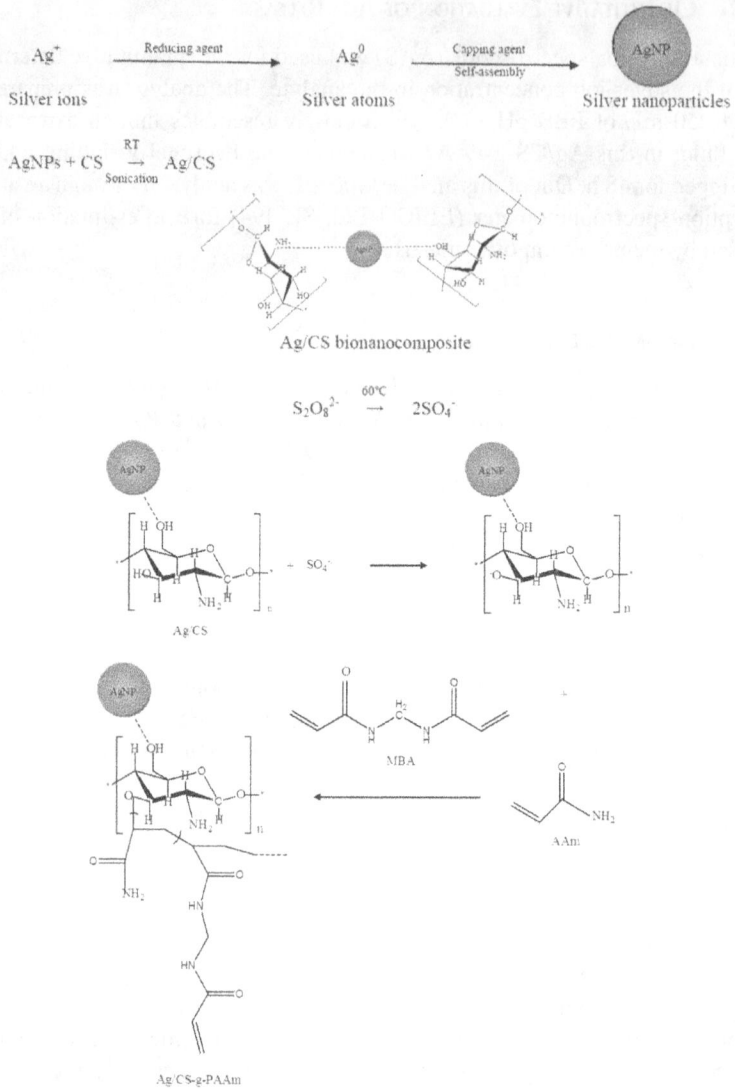

$$\text{Ag}^+ \xrightarrow{\text{Reducing agent}} \text{Ag}^0 \xrightarrow[\text{Self-assembly}]{\text{Capping agent}} \text{AgNP}$$

Silver ions   Silver atoms   Silver nanoparticles

$$\text{AgNPs} + \text{CS} \xrightarrow[\text{Sonication}]{\text{RT}} \text{Ag/CS}$$

Ag/CS bionanocomposite

$$\text{S}_2\text{O}_8^{2-} \xrightarrow{60^\circ\text{C}} 2\text{SO}_4^-$$

Ag/CS

MBA

AAm

Ag/CS-g-PAAm

FIGURE 5.1   Schematic representation of the chief reactions involved in the synthesis of Ag/CS-g-PAAm nanocomposite hydrogel.

5.3.1   FTIR ELUCIDATION

5.3.1.1   FTIR Study of the Silver Nanoparticles

In Figure 5.2, FTIR spectrum of silver nanoparticles R9, illustrated on top and labeled (AgNPs R9), elucidates the active molecules present in the aqueous rhizome extract of *Curcuma longa* that are involved in the bio-reduction of silver

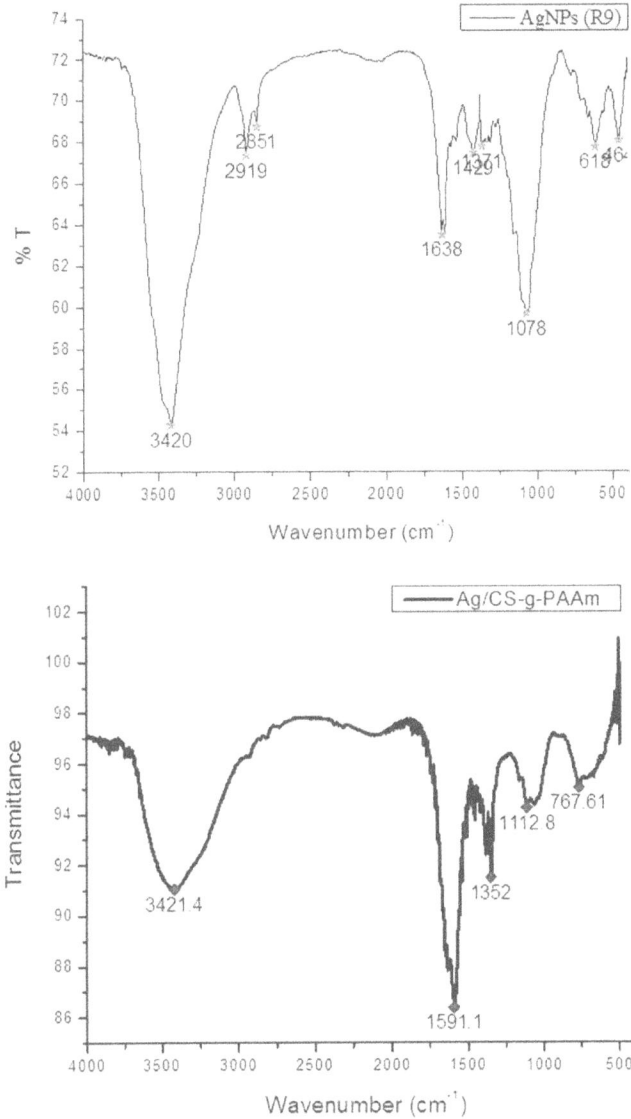

FIGURE 5.2 FTIR spectra (%Transmittance/Wavenumber) of silver nanoparticles by *Curcumalonga* and Ag/CS-g-PAAm nanocomposite.

ions and stabilizes the nanoparticles thereafter. The spectrum contains chief transmission peaks at 3420, 2919, 2851, 1638, 1429, 1371, 1078, 618, and 464 cm^{-1}. The observed sharp –OH and –CH stretches (3420 and 2919 cm^{-1}, respectively) are characteristic of sugars adsorbed on the surface of silver nanoparticles [22], which readily function as both reducing and stabilizing moieties [23, 24]. The strong peak at 3420 cm^{-1} may also be attributed to stretching of –NH amine group overlapping with –OH frequency. The peak at 1638 cm^{-1} is possibly due to the overlapping stretching vibrations of –C=C– alkene as well as –C=O carbonyl character of carboxyl group [25]. The bands at the frequency region 1429 and 1371 cm^{-1} are highly mixed. The peak at 1429 cm^{-1} may be ascribed to –C–N stretching modes of the amine [26], also it can be ascribed to the mixed in plane bending vibrations around aliphatic δ CC–C, δ CC=O and in plane bending vibrations around aromatic δ CC–H of keto and enol configurations and stretching vibrations around aromatic ν CC bonds of keto and enolic form of curcumin [27]. The band at 1371 cm^{-1} may be attributed to the pure in plane –C–H vibrations of aromatic rings. The prominent absorption peak located at 1078 cm^{-1} can be assigned as the absorption peak of stretching vibrations of –C–O–C or –C–O groups [28]. Whereas the band present at 616 cm^{-1} is probably due to –N–H bending vibrations [29]. The double bonds such as –C=C– or functional groups –C–O–C and –C–O derived from heterocyclic compounds like alkaloid or flavones, also the amide bonds of the protein residues and –NH amine groups of amino acids that are present in the rhizome extract are the capping ligands of the nanoparticles. The peak at 464 cm^{-1} can be related to bonding of nanoparticles with oxygen from hydroxyl groups of the compounds present in the aqueous extract [30].

5.3.1.2   FTIR Study of the Ag/CS-g-PAAm Nanocomposite

In Figure 5.2, in the FTIR spectrum of the Ag/CS-g-PAAm nanocomposite hydrogel, two bands peak at 3421 and 1591 cm^{-1} correspond to the primary amine and amide stretching vibrations respectively. The very sharp characteristic band at 1591 cm^{-1} is due to the –C=O asymmetric stretching. Absorption of the alcoholic –OH stretching appears in the broad range of 2550–3500 cm^{-1} [20]. The bending vibrations between 1591 and 1112 cm^{-1} intensify, indicating possible interaction between silver nanoparticles and amino group of CS.

5.3.2   Characteristic Study of the Ag/CS-g-PAAm Nanocomposite Hydrogel Super Absorbent Polymeric (SAP) Material

SEM micrograph of the nanocomposite biomaterial in Figure 5.3 displays nanoparticles uniformly dispersed in the polymer matrix. The surface of the developed Ag/CS-g-PAAm nanocomposite hydrogel is smooth. The interaction between the lone pair of electrons at the amino group as well as on the hydroxyl group of CS and the partial positive charge developed at the surface of the silver nanoparticles due to slight electron drift [31] effectively stabilizes the silver nanoparticles and prevents agglomeration.

The pattern of XRD spectrum in Figure 5.3 indicates an amorphous nature of the nanocomposite material. The broad peak or amorphous halo appearing at 2θ ~15–35° is caused by a mix of ordered and disordered structure arising due to the spacing of individual polymer chains belonging to the amorphous phase of the nanocomposite. This amorphous halo is broader in comparison to that of only

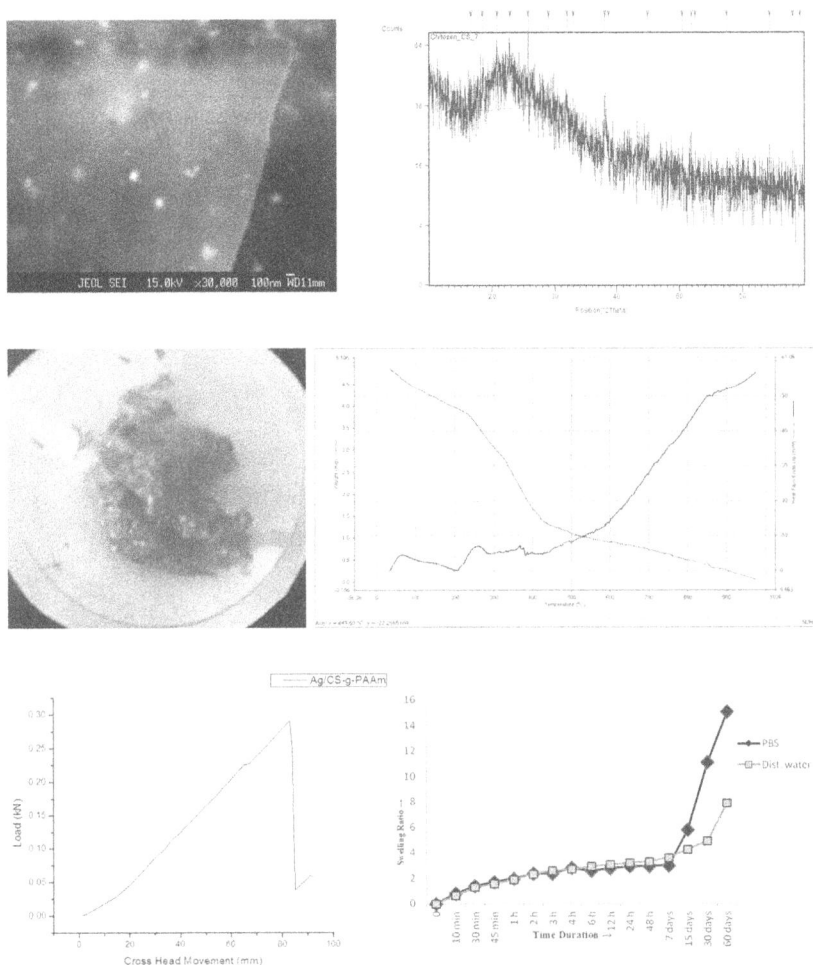

FIGURE 5.3 Illustration of the morphological and physico-chemical properties of the Ag/CS-g-PAAm nanocomposite hydrogel through SEM micrograph; XRD spectrum; visual photograph; TG/DTA curve; graph of the higher stress at maximum load, modulus and compression at break; and plotted data chart of swelling ratio (*SR*) of the Ag/CS-g-PAAm nanocomposite hydrogel vs. time.

CS biopolymer [32], which suggests the possibility of silver nanoparticles being embedded in between the chains of the grafted polymers.

The enthalpy change of the Ag/CS-g-PAAm nanocomposite hydrogel with respect to temperature and time was investigated through DSC. The thermal properties of the nanocomposite material were determined and the nanobiomaterial to decompose with melting at ~220°C. The curve indicated pure substance with endothermic process.

The TG/DTA curve of the Ag/CS-g-PAAm nanocomposite is also presented in Figure 5.3, where the TG thermogram reveals the initial temperature, T_i, and the final temperature, T_f, of the thermal-degradation process, which can be observed as ~220°C and ~420°C respectively. The DTA curve demonstrates that major mass change and probably the thermo-oxidative process of decomposition is at ~520°C. It is apparent that dispersing silver nanoparticles in CS matrix followed by graft copolymerization for the development of Ag/CS-g-PAAm nanocomposite material imparts better thermal characteristics compared to only CS biopolymer and Ag/CS bionanocomposite material [32].

5.3.3   MECHANICAL PROPERTY

According to earlier reports, reinforcement of CS matrix with nano-sized fillers promotes rigidity and better mechanical properties on several parameters due to the phenomenon of intense interactions between the reinforced particles and the matrix [33–35]. This fact is also supported by the presented data of higher stress sustained by Ag/CS-g-PAAm nanocomposite hydrogel, in Figure 5.3, which has achieved fairly good tensile strength, even much more than Ag/CS bionanocomposite because of graft copolymerization.

5.3.4   SWELLING BEHAVIOR

The swelling capacity of a nanocomposite material plays an important role in the antibacterial activity, wound healing capacity, and for various biomedical applications due to their high water/solvent holding capacity. They can further absorb moderate to fair amount of various wound exudates through swelling which intensifies fast healing. SR is the ratio of increase in swollen weight ($W_t - W_d$) to the weight of dried nanocomposite hydrogel (W_d). The SR of the Ag/CS-g-PAAm nanocomposite with time has been plotted and the graph illustrated in Figure 5.3. The chief intention in this work was to develop a self-sterilizing nanocomposite with optimum solvent absorbing property for biomedical purpose to be used as a biomaterial. It is clear from the presented data that the objective of fair solvent uptake has been achieved.

5.3.5   QUANTITATIVE EVALUATION OF AG⁺ RELEASE

AAS technique was used for the quantitative determination of the silver ion concentration in the analyte, i.e. in PBS pH ~7.0, closely resembling human extracellular body fluid. The characteristic nature of the Ag/CS-g-PAAm nanocomposite material is an Ag^+ emitter in biological and aqueous environment. It was found that Ag^+ present in 100 mL of the analyte solution taken for the test was <0.1 ppm. It is considered that steady release of Ag^+ from this Ag/CS-g-PAAm nanocomposite hydrogel is still at a level sufficient enough to render self-sterilizing property which is evident through the results of antimicrobial analysis. The Ag^+ biocide and its release increases with the amount of silver nanoparticles present in the specimen and it is also reasonable to believe that Ag^+ release will be influenced by the equilibrium swelling characteristics of the nanocomposite specimen [36–38].

TABLE 5.1
Inhibition in (mm) by the 6 mm-Piece of Safe Bioactive Ag/CS-g-PAAm Nanocomposite Hydrogel against Selective Microbial Strains

Name of the Micro-Organism	Ag/CS-g-PAAm	Reference Biomaterial (Disc-Shaped Catheter Piece)
Pseudomonas aeruginosa	21	Nil
Escherichia coli	33	Nil
Staphylococcus aureus	18	Nil

5.3.6 ANTIMICROBIAL ASSAY

Details of the result obtained are listed in Table 5.1, of the developed Ag/CS-g-PAAm nanocomposite hydrogel, assayed for antimicrobial activity against *P. aeruginosa* (Gram negative), *E. coli* (Gram negative) and *S. aureus* (Gram positive) causing majority of the biomedical implant related infection. It has been observed in the present study that the effect was well pronounced against Gram-negative bacteria which contain only a thin peptidoglycan layer of 2~3 nm between the cytoplasmic membrane and the outer membrane, and also against Gram-positive bacteria which lack the outer membrane but have a peptidoglycan layer of about 30-nm thickness. Silver nanoparticles present in the material when exposed to aqueous environment releases silver ions (Ag^+) which manifest antibacterial activity by anchoring to and penetrating the bacterial cell wall and modulating cellular signaling, finally causes cell lysis by rupturing the cytoplasmic membrane [39].

5.3.7 *IN VITRO* BLOOD COMPATIBILITY

The Ag/CS-g-PAAm nanocomposite hydrogel biomaterial did not cause any noticeable hemolysis of the blood sample to which it was exposed, because a sample is classified as being hemolyzed if the A_{414} reading exceeds a value of 0.2 [40]. The readings of the optical density assessment are summarized in Table 5.2. When the cell count changes (leukocytes, erythrocytes and platelets) in reference are compared with the test sample, none of the parameters showed significant

TABLE 5.2
Optical Density (OD) Reading

Tube	Sample	OD
1	Ag/CS-g-PAAm	0.089
2	Control	0.139
3	Reference	0.136

TABLE 5.3

Blood Cell Counts of the Samples

Tube	Sample	Leukocytes (WBC)	Erythrocytes (RBC)	Platelets (PLT)	Hemoglobin (HGB)
1	Ag/CS-g-PAAm	$6.4 \times 10^3/\mu L$	$4.13 \times 10^6/\mu L$	$153 \times 10^3/\mu L$	13.1 g/dL
2	Control	$6.4 \times 10^3/\mu L$	$4.06 \times 10^6/\mu L$	$161 \times 10^3/\mu L$	12.9 g/dL
3	Reference	$6.2 \times 10^3/\mu L$	$4.12 \times 10^6/\mu L$	$162 \times 10^3/\mu L$	13.2 g/dL

variation as an effect of material contact with blood. Results of the analysis are listed in Table 5.3. The nanocomposite is possibly possessed with factors reducing hemolysis while having good blood compatibility [41].

5.4   PRINCIPLE OF APPLICATION

For dealing with BAI, one of the most alarming problems in biomedical engineering, we have overcome with an innovative technique to develop this novel safe and bioactive self-sterilizing Ag/CS-g-PAAm nanocomposite hydrogel that can indubitably be applicable as a catheter material or in the essential modified form of other required biomaterials. It is cost efficient, expected to be nontoxic and biocompatible, for a broad range of wide spread use as nanobiomaterial in the sector of medicine, biomedical and life science. Application of the biofunctionalized nanobiomaterial encompasses hospitals, clinics, personal health care systems, etc.

5.5   CONCLUSION

Thus, this silver-reinforced nanocomposite hydrogel biomaterial was synthesized through phytofabrication route involving the rhizome of *Curcuma longa* (turmeric). Curcumin is the most bioactive ingredient of this part of the plant, which is well known for ages to be endowed with immense inherent medicinal properties. Silver ions were reduced and nanoparticles were formed when exposed to turmeric extract containing the plant organic matter and biomolecules. From this study it was concluded that active aqueous silver ions were released from the nanocomposite when exposed to aqueous environment administrating enhanced antimicrobial activity. Therefore, silver-reinforced CS-g-PAAm nanocomposite hydrogel expresses antimicrobial characteristics in comparison to only CS-g-PAAm hydrogel, which does not possess any such quality [20]. Connecting nature to nanotechnology for dealing with one of the increasingly growing problems in medicine and biomedical engineering, we have successfully innovated a novel self-sterilizing Ag/CS-g-PAAm nanocomposite hydrogel that can be indubitably effectively applicable as a catheter material or in the essential form of other crucial biomaterials. It is cost efficient, expected to be nontoxic and biocompatible.

REFERENCES

1. R. O. Darouiche, N. Engl. J. Med., 350 (2004) 1422.
2. A. Trampuz and W. Zimmerli, Curr. Opin. Investig. Drugs, 6 (2005) 185.
3. I. Fundeanu, H. C. Van der Mei, A. J. Schouten and H. J. Busscher, Colloids Surf. B Biointerfaces, 64 (2008) 297.
4. J. Parvizi, V. Antoci Jr., N. J. Hickok and I. M. Shapiro, Exp. Rev. Med. Dev., 4 (2007) 55.
5. S. Samuggam, S. V. Chinni, P. Mutusamy, S. C. B. Gopinath, P. Anbu, V. Venugopal, L. V. Reddy and B. Enugutti, Molecules, 26 (2021) 2681.
6. B. D. Ratner, Polym. Int., 56 (2007) 1183.
7. E. M. Hetrick and M. H. Schoenfisch, Chem. Soc. Rev., 35 (2006) 780–789.
8. M. Catauro, F. de Gaetano and A. Marotta, J. Mater. Sci., 15 (2004) 831.
9. D. C. Tien, K. H. Tseng, C. Y. Liao and T. T. Tsung, J. Alloys Compd., 473 (2009) 298.
10. J. H. Crabtree, R. J. Burchette, R. A. Siddiqi, I. T. Huen, L. L. Handott and A. Fishman, Perit. Dial. Int., 23 (2003) 368.
11. J. J. Castellano, S. M. Shafii, F. Ko, G. Donate, T. E. Wright, R. J. Mannari, W. G. Payne, D. J. Smith and M. C. Robson, Int. Wound J., 4 (2007) 114.
12. J. S. Kim, E. Kuk, K. N. Yu, J.-H. Kim, S. J. Park, H. J. Lee, S. H. Kim, Y. K. Park, Y. H. Park, C. Y. Hwang, Y. K. Kim, Y.-S. Lee, D. H. Jeong and M. H. Cho, Nanomedicine, 3 (2007) 95.
13. H. Kong and J. Jang, Langmuir, 24 (2008) 2051.
14. P. N. Ravindran, K. N. Babu and K. Sivaraman, Turmeric: The Genus Curcuma (Medicinal and Aromatic Plants – Industrial Profiles) (1st ed.), CRC, Boca Raton, London, & New York (2007).
15. P. Dwivedi, D. Tiwary, S. S. Narvi, R. P. Tewari and K. P. Shukla, Lett. Appl. Nanobiosci., 9 (2020) 1485–1493.
16. B. A. Cheba, Procedia Manuf., 46 (2020) 652–658.
17. M. Maddaloni, I. Vassalini and I. Alessandri, Sus. Chem., 1 (2020) 325–344.
18. G. Mohammadkhani, S. K. Ramamoorthy, K. H. Adolfsson, A. Mahboubi, M. Hakkarainen and A. Zamani, Polymers, 13 (2021) 2121.
19. C. X. Xie, Y. J. Feng, W. P. Cao, H. K. Teng, J. F. Li and Z. Y. Lu, J. Appl. Polym. Sci., 111 (2009) 2527.
20. A. Pourjavadi and G. R. Mahdavinia, Turk. J. Chem., 30 (2006) 595–608.
21. R. Kumar and H. Munstedt, Polym. Int., 54 (2005) 1180.
22. G. Von White II, P. Kerscher, R. M. Brown, J. D. Morella, W. McAllister, D. Dean and C. L. Kitchens, J. Nanomater., Article ID 730746 (2012) 1–12.
23. J. Liu, M. Anand and C. B. Roberts, Langmuir, 22 (2006) 3964–3971.
24. S. K. Mehta, S. Chaudhary and M. Gradzielski, J. Coll. Int. Sci., 343 (2010) 447–453.
25. K. Shameli, M. B. Ahmad, A. Zamanian, P. Sangpour, P. Shabanzadeh, Y. Abdollahi and M. Zargar, Int. J. Nanomed., 7 (2012) 5603–5610.
26. S. M. Ghoreishi, M. Behpour and M. Khayatkashani, Physica E., 44 (2011) 97–104.
27. T. M. Kolev, E. A. Velcheva, B. A. Stambolyska and M. Speteller, Int. J. Quant. Chem., 102 (2005) 1069.
28. J. Huang, Q. Li, D. Sun, Y. Lu, Y. Su, X. Yang, et al., Nanotechnology, 18 (2007) 105104–105114.
29. K. Gopinath, S. Gowri, V. Karthika and A. Arumugam, J, Nanostruc. Chem., 4 (2014) 1–11.
30. K. Shameli, M. B. Ahmad, S. D. Jazayeri, P. Shabanzadeh, P. Sangpour, H. Jahangirian, et al. Chem. Cent. J., 6 (2012) 73–95.
31. S. W. Kang, K. Char and Y. S. Kang, Chem. Mater., 20 (2008) 1308–1311.

32. L. Balau, G. Lisa, M. I. Popa, V. Tura and V. Melnig, Cent. Eur. J. Chem., 2 (2004) 638–647.
33. K. A. M. Amin and M. in het Panhuis, Polymers, 4 (2012) 590–599.
34. S. Azizi, M. B. Ahmad, M. Z. Hussein, N. A. Ibrahim and F. Namvar, Int. J. Nanomed., 9 (2014) 1909–1917.
35. P. Cazón and M. Vázquez, Environ. Chem. Lett., 18 (2020) 257–267.
36. R. P. S. P. L. Williams, P. J. Doherty, D. G. Vince, G. J. Grashoff and D. F. Williams, Crit. Rev. Biocompat, 5 (1989) 221–223.
37. T. J. Berger, J. A. Spadaro, S. E. Chapin and R. O. Becher, Antimicrob. Agents Chemother., 9 (1976) 357–358.
38. R. M. Slawson, M. I. Vandyke, H. Lee and J. T. Trevors, Plasmid, 27 (1992) 72–77.
39. S. Shrivastava, T. Bera, A. Roy, G. Singh, P. Ramchandrarao and D. Dash, Nanotechnology, 18 (2007) 225103–225111.
40. M. B. Kirschner, J. J. B. Edelman, S. C. H. Kao, M. P. Vallely, N. V. Zandwijk and G. Reid, Front. Genet., 4 (2013) 94–101.
41. L. A. Harker, B. D. Ratner and P. Didisheim, Cardiovasc. Pathol., 2, Suppl 3 (1993).

6 Bioactive Silver/ Chitosan/Polyvinyl Chloride (Ag/CS/PVC) Nanocomposite Blend
Phytomass Enabled

6.1 INTRODUCTION

The field of manufacturing technology has totally been renovated with the emergence of nanotechnology, whereas mergence of nanoscience with material science, biology, biotechnology and medicine has completely novelized various arrays of biomedical engineering [1, 2]. Advancement in the field of nanomaterial manufacturing for biomedical engineering [3], largely depend on the ability to synthesize nanoparticles in an eco-friendly manner without using any toxic chemicals in the protocol [4–6]. In the present work, bio-reduction and formation of silver nanoparticles have been achieved, by using the bark of the plant *Terminalia arjuna*, which is commonly known as arjuna bark or arjun, one of the best cardioprotective agents.

Since time immemorial according to classical Ayurvedic medicine and Unani description, *Terminalia arjuna* especially, its bark, is considered to be an excellent astringent, anticoagulant, antihypertensive, antithrombotic, cardiac stimulant, hemostatic, lithontriptic, rejuvenative tonic, having antiviral, antifungal and antibacterial properties. It is also helpful in hypertension, ischemic heart disease (IHD), congestive cardiac failure and possesses wide ranging therapeutic properties having the potential to treat numerous medical conditions, especially those pertaining to the heart and circulatory system [7]. Even today, modern practitioners consider the arjuna bark as a cardiac tonic; a decoction and the preparations made by it are reputed to have a market-stimulant action on the heart. The bark endorses effectual functioning of the cardiac muscles and is thus for centuries being used for maintaining the health of the cardiovascular system. The bark contains small amounts of several phytochemicals, specifically a crystalline compound Arjunine, also Arjunetin, and a new acid called Arjunic acid [8].

In this investigation, for the first time, the silver nanoparticles fabricated with the aid of the arjuna bark extract have been incorporated into a chitosan (CS)/polyvinyl chloride (PVC) blended matrix to develop a novel silver/chitosan/

DOI: 10.1201/9781003217343-6

polyvinyl chloride (Ag/CS/PVC) antimicrobial self-sterilizing nanocomposite blend biomaterial or nanobiomaterial. The blending of PVC with CS was enabled through simultaneous casting of their separate solutions in 2:1 proportion of CS and PVC in suitable solvents [9], glutaraldehyde (GA) was used to ensure cross-linking between the two polymers [10, 11]. CS, a natural biopolymer composed of poly(β-(1-4)-2-amino-2-deoxy-D-glucose), is one of the structural polysaccha-ride which is abundantly available in nature after cellulose [12–14]. CS interacts very easily with bacterium and binds to DNA, glycosaminoglycans and most of the proteins, thereby enhancing the antimicrobial effect of silver nanoparticles [15–17], together with effectively stabilizing the nanoparticles in the matrix and preventing agglomeration [18].

PVC is one of the most commonly used synthetic polymers in biomedical applications, especially as a catheter material for the circulatory system [19]. Earlier studies shed light on the underappreciated and significant risks of bioma-terials associated infections (BAI) through peripheral IVs as well as other car-diovascular catheters composed of PVC [20]. A new study has found that more than one in ten catheter-related bloodstream infections due to *Staphylococcus aureus* in hospitalized adults are caused by infected peripheral venous catheters [21] also known as peripheral IVs composed of PVC [22, 23]. Peripheral IVs as well as other cardiovascular catheters are the ubiquitous aspect of hospital patient care. It is indicative of immensely important need for initiation of interventions to combat the risk and warfare of BAI.

Interventions to mitigate the risk of BAI may be achieved up to a great extent if PVC could be converted into value-added self-sterilizing biomaterial. PVC has been scarcely introduced as a candidate in a chemically modified form [24]. One of the trials was conducted by Rinaudo to modify PVC with quaternary ammo-nium salts through physical mixing with slightly cross-linked PVC [14]. But the chemical modification of PVC materials is still relatively low this of course is due to the sensitive nature of PVC terminal double bonds toward thermal effects during the different chemical modifications undertaken on PVC-based materi-als including cross-linking. In recent years, much attention has been focused on the biomedical applications of synthetic polymers including poly(vinyl alcohol), poly(vinyl acetate), poly(methyl methacrylate), etc, by incorporating a second polymer component, viz. CS [2–27]. An improvement in properties of the blend system can be achieved. This can be attributed to the good biocompatibility, antibacterial properties, appropriate biodegradability, excellent physico-chem-ical properties, and its commercial availability at relatively low cost. Recently, researchers have reported application of polymer blends for biomedical appli-cations because they provide an efficient way to fulfill new requirements for material properties. One of the widely used polymers to be incorporated into CS is poly(vinyl alcohol), it is suggested that enhancement in tensile strength of the blend may be due to the hydrogen bonding between hydroxyl groups of PVA and amine or hydroxyl groups of CS [28, 29]. However, polymers have also been reported to be immiscible in several cases, prepared hydrogels of CS/PVA blends indicated that CS was enriched on the surface of gel membranes.

This implied occurrence of a phase separation in previously synthesized CS/ PVA blends. Moreover, CS/PVC blend has not yet been explored for potential biomedical engineering applications.

So in our study we have approached blending of PVC with CS, using GA as a cross-linker to enhance the CS/PVC compatibility and its ensuing properties for biomedical use. The obtained blend was found to have reasonable extent of compatibility between their components. Such compatibility mainly depends on the manner in which the components of the nanocomposite have been blended together with each other. In addition, we have also dispersed the phytofabricated silver nanoparticles to this cross-linked copolymer matrix, which renders the developed nanocomposite self-sterilizing, with overwhelming antibacterial response for effective use as a catheter material to overcome BAI. This work also presents a facile approach for developing the novel nanobiomaterial. Ag/ CS nanocomposite was first prepared through phytomass intervened value-added conversion and then transformed to Ag/CS-PVC nanocomposite blend. Relevant techniques have been used for characterization.

6.2   EXPERIMENTAL

6.2.1   MATERIALS

CS (degree of deacetylation: 79%, molecular mass: 500,000 g/mol) purchased from Sea Foods (Cochin), India; PVC (molecular weight 10,000) purchased from Sigma Aldrich; glutaraldehyde (GA), tetrahydrofuran (THF), acetic acid glacial (extra pure) and nutrient agar were purchased from Thomas Baker (Chemical) Pvt. Ltd. India. Silver nanoparticle suspension from reaction mixture (R2), phytofabricated by the bark of the plant *Terminalia arjuna* (arjuna bark). All the chemicals and reagents were of analytical grade; therefore, were used without further purification. Bacterial strains, *Pseudomonas aeruginosa* (Gram negative), *Escherichia coli* (Gram negative), *Staphylococcus epidermidis* (Gram positive), *S. aureus* (Gram positive) and *Bacillus subtilis* (Gram positive), were obtained from the culture bank of Microbiology Department, Sam Higginbottom Institute of Agriculture, Technology and Sciences, Allahabad, India. Deionized water was also used for preparation of solutions.

6.2.2   SYNTHESIS – SYNTHESIS OF AG/CS/PVC NANOCOMPOSITE BLEND

The residue of silver nanoparticles (R2) obtained from the suspension of silver nanoparticles (RM2), phytofabricated by arjuna bark (experimental details referred and discussed in Chapter 3), was dispersed in 20 mL of CS solution (2% [w/v] in 1% [v/v] acetic acid) and sonicated for 10 min. This Ag/CS nanocomposite solution (NC) prepared was properly kept for further use. NC solution (20 mL) and PVC solution (10 mL) (1% [w/v] in THF) were taken. The two solutions in the ratio 2:1 were poured simultaneously in another beaker portion-wise while stirring and 1 mL of 25% GA for proper cross-linking was added to the blended mixture. This blend was sonicated for further 30 min then left to stand at room temperature. The obtained product, Ag/CS/PVC nanocomposite blend, was collected and dried in vacuum at 37°C for 24 h [9].

6.2.3 FTIR SPECTROSCOPY

6.2.3.1 FTIR Study of the Silver Nanoparticles

The dried R2 residue of silver nanoparticles, phytofabricated by arjuna bark extract, was characterized using Fourier transformed infrared spectroscopy (FTIR). The FTIR analysis was carried out to have an overview of the possible bio-molecules responsible for the reduction of silver ions (Ag^+) and identify the functional groups stabilizing the silver nanoparticles. This also gives a rough account of the beneficial compounds of the bark extract adsorbed on the surface of the nanoparticles acting as capping agents. The isolated and purified residue of silver nanoparticles free from other loosely bound bio-organic molecules present in the suspension were received after repeated centrifugation. FTIR spectrum of these silver nanoparticles was recorded over the range of (500–4000) cm^{-1} with (PerkinElmer) spectrophotometer for studying the chemical characteristics.

6.2.3.2 FTIR Study of the Ag/CS/PVC Nanocomposite

FTIR spectra of CS, PVC and Ag/CS/PVC were recorded over the range of (500–4000) cm^{-1} with (PerkinElmer) spectrophotometer to assess the chemical properties of the of the nanocomposite blend material.

6.2.4 CHARACTERIZATION OF AG/CS/PVC NANOCOMPOSITE

The synthesized Ag/CS/PVC nanocomposite blend was characterized using scanning electron microscopy (SEM), X-ray diffraction (XRD), differential scanning calorimetry (DSC), thermo-gravimetric/differential thermal analysis (TG/DTA). DSC analysis was done (using Metler Toledo DSC 25) and TG/DTA of the biomaterial were carried out using (Perkin Elmer, Pyris Diamond, Tg-DTA high temp 115V) TGA instrument. The structure of the nanocomposite material was studied using XRD (XRD, Philips, Xpert, Cu Kα) at a scanning speed of 3°/min. SEM analysis was done (through JEOL JXA 8100) for morphological study of the biomaterial.

6.2.5 SWELLING PARAMETERS

The completely dried, pre-weighed Ag/CS/PVC nanocomposite samples, weighing 0.0109 and 0.0033 g, were equilibrated in 250 mL of phosphate buffer saline solution (PBS) at pH ~ 7.0 and 250 mL of deionized water respectively, at room temperature. After immersion in PBS and deionized water separately, the nanocomposite samples were removed at different time intervals and blotted with filter paper to remove excess solvent from the surface. The solvent uptake by the nanocomposite samples at time intervals were measured using an analytical electronic balance. The swelling ratio (SR) was calculated using the following equation,

$$\text{Swelling ratio } (SR) = ([W_t - W_d]/W_d)$$

where, W_t denotes weight of swollen material at time (t) and W_d denotes weight of dry material (before swelling).

6.2.6 QUANTITATIVE EVALUATION OF AG⁺ RELEASE

Atomic absorption spectroscopy (AAS) was used for the quantitative determination of the silver ion concentration in the analyte. The analyte was prepared by taking 250 mL of PBS pH ~ 7.0, which closely resembles human extracellular body fluid; in this Ag/CS/PVC nanomaterial weighing 16 mg was dipped for 48 h. 100 mL of this analyte was analyzed through atomic absorption spectrophotometer (ELICO Ltd. SL 194) for the estimation of Ag⁺ released by the nanocomposite material.

6.2.7 MECHANICAL TESTING

Mechanical properties of Ag/CS/PVC nanocomposite material was measured using (INSTRON 1195, Universal Testing Machine UTM, Buckinghamshire, England) running at a crosshead speed of 0.5 mm/min. The sample material was cut into 20×50-mm size and the gauge length was about 10 mm. The mechanical testing was done using 20 kN load cell for tensile parameters, maximum stress and % compression at break, which were measured and plotted.

6.2.8 ANTIMICROBIAL ASSAY

The Ag/CS/PVC nanocomposite material was assayed for antimicrobial activity against *P. aeruginosa* (Gram negative), *E. coli* (Gram negative), *S. epidermidis* (Gram positive), *S. aureus* (Gram positive) and *B. subtilis* (Gram positive). Disc diffusion method [30] was used to find out the standard zone of inhibition (ZOI). The nanomaterial was cut into disc shape having ~5-mm diameter, sterilized by UV radiation for 30 min and placed on different cultured agar plates. Nutrient agar was used as culture media and inoculated with 1000 μL of bacterial organism containing broth. These plates were incubated at 37°C for 48 h. The plates were then examined for evidence of ZOI, which appear as a clear area around the disc. A meter ruler was used to measure the diameter of such ZOI.

6.2.9 *IN VITRO* BLOOD COMPATIBILITY TEST

Five tubes were taken into consideration and marked as '1, 2, 3, 4, 5'; containing 1 mL each of anticoagulated human blood. The test sample, a piece of Ag/CS/PVC nanocomposite material weighing 0.0098 g, was then exposed to the anticoagu-lated human blood by suspending the sample material in tube (1) with the help of a sterilized thread. In tubes (2) and (3), 5×5-mm sample pieces, of only PVC film weighing 0.0069 g and only CS film weighing 0.0010 g, were suspended respec-tively with the help of sterilized threads. In tube (4), a strand of sterilized thread alone was suspended for control, whereas tube (5) was used for reference without any material and thread while containing only 1 mL of anticoagulated human blood. After observation and exposure of the samples for 1 h; they were taken out.

Blood cell counts (leukocytes, erythrocytes and platelets) were measured using Beckman Coulter Counter (HMX) hematology analyzer. The degree of hemoly-sis was estimated by spectrophotometry through optical density measurement of blood plasma against normal saline, at wavelengths scanning from 350 to 650 nm, with the help of the instrument (Erba CHEM-5 Plus v2). The level of hemolysis was determined based on the absorbance peak of free hemoglobin at 414 nm.

6.3   RESULTS AND DISCUSSION

The general mechanism for the development of Ag/CS/PVC nanocomposite blend can be represented through the following schematic reactions given in Figure 6.1.

6.3.1   FTIR ELUCIDATION

6.3.1.1   FTIR Study of the Silver Nanoparticles

In Figure 6.2, FTIR spectrum of the R2 silver nanoparticles (AgNPs R2) elu-cidates the active molecules present in the aqueous arjuna bark extract that are involved in the bio-reduction of silver ions and stabilizes the nanoparticles there-after. The spectrum contains chief transmission peaks at 3415, 2919, 2850, 1618, 1399, 1373, 1236, 1077, 616 and 484 cm^{-1}. The observed sharp –OH and –CH stretches (3415 and 2919 cm^{-1} respectively) indicate sugars adsorbed on the sur-face of silver nanoparticles [31], which readily function as reducing and stabiliz-ing agents [32, 33]. The strong absorption band at 3415 cm^{-1} is characteristic stretching vibration frequencies of both alcoholic –OH and –NH amine groups. The peaks between 2919 and 2850 cm^{-1} are typically for a –C–H symmetrical vibration of saturated hydrocarbons [34]. The peak at 1618 cm^{-1} is possibly due to the overlapping stretching vibrations of –C=C– alkene as well as –C=O carbonyl character of carboxyl group [35]. The bands in the frequency region 1399–1373 cm^{-1} are highly mixed, may probably be due to –C–N stretching modes of the amine, while the band at 1236 cm^{-1} may be attributed to the pure in plane –C–H vibrations of aromatic rings [36]. The absorption peak at 1077 cm^{-1} can be assigned for the stretching vibrations of –C–O–C or –C–O groups [37]. Whereas the band at 616 cm^{-1} position is probably because of –N–H bending vibrations [38]. The double bonds such as –C=C– or functional groups like the –C–O–C and –C–O derived from heterocyclic compounds like alkaloid or flavones, also the amide bonds of the protein residues and –NH amine groups of amino acids

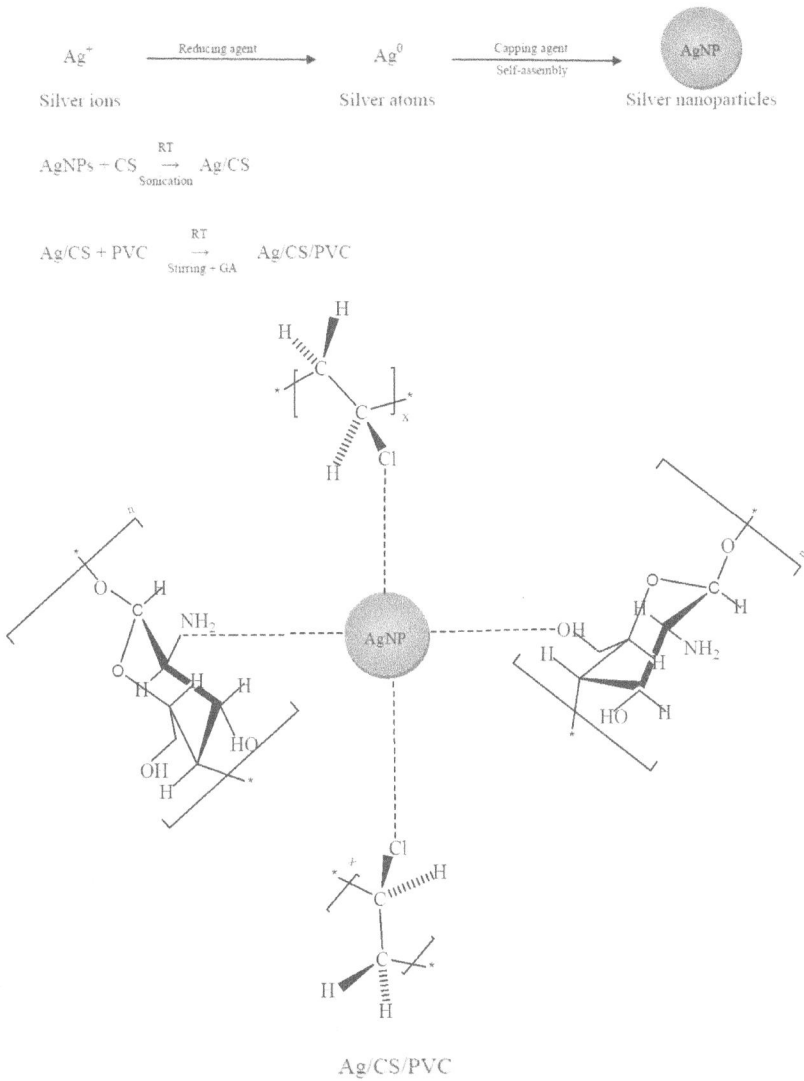

FIGURE 6.1   Schematic representation of the chief reactions involved in the synthesis of Ag/CS/PVC nanocomposite blend as bioactive nanobiomaterial.

that are present in the bark extract are the capping ligands for the nanoparticles. The peak at 484 cm^{-1} can be related to bonding of nanoparticles with oxygen from hydroxyl groups of the compounds present in the aqueous arjuna bark extract [39].

6.3.1.2   FTIR Study of the Ag/CS/PVC Nanocomposite

Changes in chemical structure are well elucidated through the FTIR spectra. In Figure 6.2, in the FTIR spectra of CS and Ag/CS/PVC nanocomposite blend, the

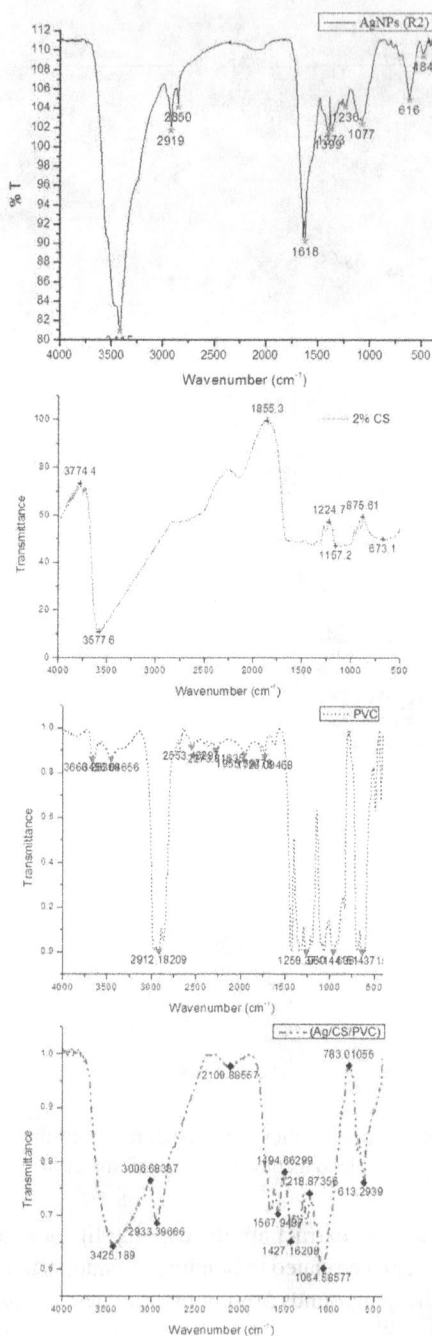

FIGURE 6.2 FTIR spectra (% Transmittance/Wavenumber [cm^{-1}]) of silver nanoparticles by arjuna bark, 2% CS, 1% PVC and Ag/CS/PVC nanocomposite blend.

peak at ~3500 cm^{-1} is more pronounced corresponding to the axial OH group of the CS molecule, or may also correspond to the primary amine stretching vibrations. The bending vibrations between 1600 and 1000 cm^{-1} intensify in the FTIR spectrum of Ag/CS/PVC than in the FTIR spectrum of CS indicating possible interaction between silver nanoparticles and amino group of CS. The broadening between 2500 cm^{-1} and 1800 cm^{-1} in the FTIR spectrum of Ag/CS/PVC denotes crosslinked blending of PVC with CS. The band at 1568 cm^{-1} is due to the free –NH$_2$ groups and the very sharp band at 1065 cm^{-1} for C–O–C bonds, which showed the characteristics of CS in the FTIR spectrum of Ag/CS/PVC. An absorption band at 2933 cm^{-1} for C–H stretching is also present in the FTIR spectrum of Ag/CS/PVC, attributed to both CS and PVC [9]. This indicates some inter-chain distribution of CS and PVC, consequently a blend matrix with reasonable extent of homogeneity, having dispersion of silver nanoparticles has been obtained.

6.3.2 CHARACTERISTIC STUDY OF THE AG/CS/PVC NANOCOMPOSITE

The structure and morphological characteristics of the developed Ag/CS/PVC nanocomposite blend material was studied through SEM. SEM micrographs in Figure 6.3 show nanoparticles dispersed in the CS/PVC copolymer matrix with minimum aggregation. The smooth homogenous mixing and optimized blending of the two polymer components in the nanobiomaterial is also vividly elucidated through the SEM micrographs.

The XRD pattern in Figure 6.3 indicates a semi-crystalline structure of the nanobiomaterial. Three diffraction peaks observed at $2\theta = 38°$, 44° and 64° correspond to the (111), (200) and (220) Miller indices crystallographic planes, respectively, of the face-centered cubic (FCC) Ag°, (JCPDS file No. 00-004-0783). The peaks obtained are not very sharp, as a result of the capped silver nanoparticles by the plant biomolecules on the surface. The existence of two broad peaks at 2θ ~11° and 24° indicates the presence of amorphous CS and PVC components. The broadening of the amorphous halo increases suggesting silver nanoparticles being embedded in between the chains of individual polymers.

The enthalpy change of the Ag/CS/PVC nanocomposite blend material with respect to temperature and time was investigated through DSC. The nanobiomaterial at ~260°C undergoes further endothermic decomposition. The TG/DTA of the Ag/CS/PVC nanocomposite blend is illustrated in Figure 6.3. The thermal properties of the nanocomposite blend appear to be complex. The thermograms are clearly revealing the thermal degradation process. The TG curve represents the slope and noticeable peak for identifying the degradation stages. While, the DTA curve demonstrates the endothermic effect and denotes very high degradation temperature, which results into depolymerization and thermo-oxidative decomposition. Therefore, it may be concluded that reinforcing silver nanoparticles in CS matrix together with simultaneous blending with PVC confers better thermal stability in comparison to only CS films according to the DSC and DTA curves [40].

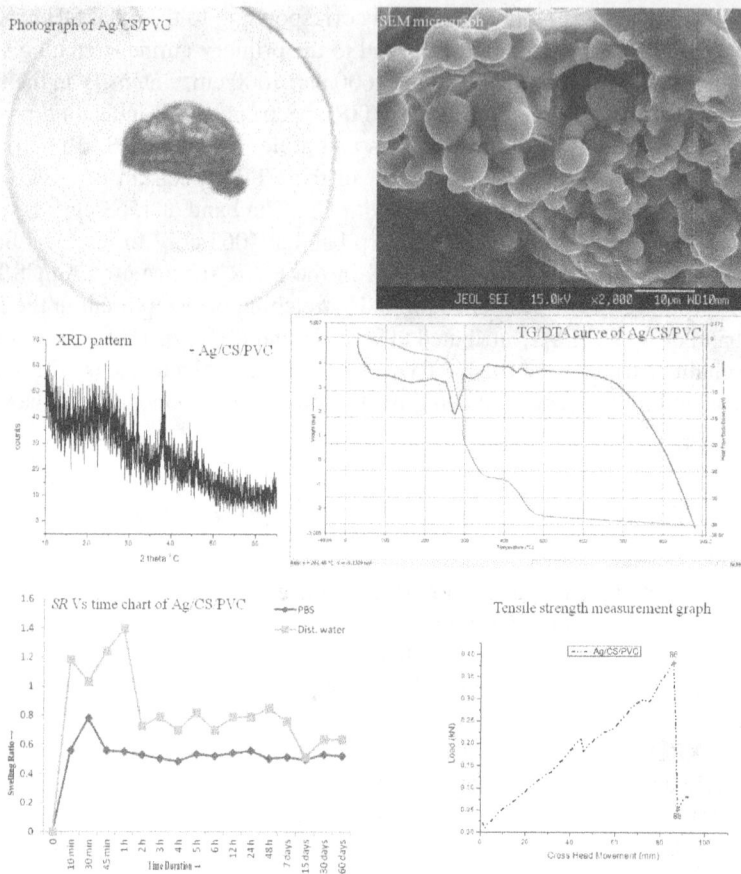

FIGURE 6.3 Illustration of the morphological and physico-chemical properties of the Ag/CS/PVC nanocomposite blend through visual photograph; SEM micrograph; XRD spectrum; TG/DTA curve; plotted data chart of *SR* vs. time and graph of the higher stress at maximum load, modulus and compression at break.

6.3.3 SWELLING BEHAVIOR

The swelling capacity of a nanocomposite material plays an important role in the antibacterial activity, wound healing capacity and for various biomedical applications due to their water/solvent holding capacity. The *SR* is the ratio of increase in swollen weight ($W_t - W_d$) to the weight of dried Ag/CS/PVC nanocomposite material (W_d). The *SR* of the Ag/CS/PVC nanocomposite against time has been plotted in the graph depicted in Figure 6.3. The higher crosslink's within the nanobiomaterial restricts the penetration of solvent for swelling. But the increase of CS content in the nanocomposite (2:1, CS: PVC) increases the swelling capacity significantly.

6.3.4 QUANTITATIVE EVALUATION OF AG⁺ RELEASE

AAS technique was used for the quantitative determination of the silver ion concentration in the analyte, i.e. in PBS pH ~ 7.0, closely resembling human extracellular body fluid. The characteristic nature of the Ag/CS/PVC nanocomposite material is an Ag^+ emitter in biological and aqueous environment. It was found that Ag^+ presents in 100 mL of the analyte solution taken for the test was less than 0.1 ppm. It is considered that steady release of Ag^+ from the developed Ag/CS/PVC nanocomposite blend is significantly low but at a level sufficient enough to render remarkable self-sterilizing property, which is evident through the antimicrobial analysis done. The Ag^+ biocide release increases with the amount of silver nanoparticles present in the specimen and will be influenced by the equilibrium swelling characteristics of the nanocomposite specimen [41–43].

6.3.5 MECHANICAL PROPERTIES

Earlier reports suggest that rigidity is promoted through reinforcement of CS matrix with nano-sized fillers, imparting enhanced mechanical properties on several parameters based on the phenomenon of intense interactions between the reinforced particles and the matrix [44–46]. This fact is being adamantly supported by the present data of Ag/CS/PVC nanocomposite blend, in Figure 6.3, which has achieved better strength in comparison to only CS, Ag/CS or even Ag/CS-g-PAAm, probably due to blending of polymers which are maintaining their phases and are able to exert their characteristic properties.

6.3.6 ANTIMICROBIAL ASSAY

Details of the result obtained are listed in Table 6.1, of the Ag/CS/PVC nanocomposite blend, assayed for antimicrobial activity against *P. aeruginosa* (Gram negative), *E. coli* (Gram negative), *S. epidermidis* (Gram positive), *S. aureus* (Gram positive) and *B. subtilis* (Gram positive).

It has been observed in the present study that the effect was well pronounced against Gram-negative bacteria which contain only a thin peptidoglycan layer

TABLE 6.1
Zone of Inhibition (ZOI) in (mm) of the Phytomass Fabricated Ag/CS/PVC Nanobiomaterial against Selective Bacterial Strains

Name of the Micro-Organism	Ag/CS/PVC
Pseudomonas aeruginosa	36.1
Escherichia coli	43
Staphylococcus epidermidis	34.8
Staphylococcus aureus	35
Bacillus subtilis	23.4

TABLE 6.2
Optical Density (OD) Values of the Samples

Tube	Sample	OD
1	Ag/CS/PVC	0.134
2	PVC film	0.213
3	CS film	0.111
4	Reference	0.136
5	Control	0.139

of 2~3 nm between the cytoplasmic membrane and the outer membrane, and slightly less pronounced against Gram-positive bacteria which lack the outer membrane but have a peptidoglycan layer of about 30-nm thickness. Silver nanoparticles when exposed to aqueous environment releases Ag^+ which binds with the thiol groups of certain amino acids and inhibits the enzymes of respiratory cycle and also interferes with the DNA replication of the micro-organism [47].

6.3.7 *In vitro* Blood Compatibility

The Ag/CS/PVC nanocomposite blend material did not cause any noticeable hemolysis of the blood sample to which it was exposed, because a sample is classified as being hemolyzed if the A_{414} reading exceeds a value of 0.2 [48]. The readings of the optical density assessment are summarized in Table 6.2. When the cell count changes (leukocytes, erythrocytes and platelets) in reference are compared with the test sample, none of the parameters showed any significant variation as an effect of nanomaterial contact with blood. Results of the analysis are listed in Table 6.3. It was concluded that Ag/CS/PVC nanocomposite is also possibly possessed with factors reducing hemolysis while having good blood compatibility [49].

TABLE 6.3
Blood Cell Counts of the Samples for the Determination of *in vitro* Blood Compatibility of the Developed Ag/CS/PVC

Tube	Sample	Leukocytes (WBC)	Erythrocytes (RBC)	Platelets (PLT)	Hemoglobin (HGB)
1	Ag/CS/PVC	$6.20 \times 10^3/\mu L$	$4.23 \times 10^6/\mu L$	$168 \times 10^3/\mu L$	13.60 g/dL
2	PVC film	$6.10 \times 10^3/\mu L$	$4.24 \times 10^6/\mu L$	$164 \times 10^3/\mu L$	13.50 g/dL
3	CS film	$6.41 \times 10^3/\mu L$	$4.32 \times 10^6/\mu L$	$173 \times 10^3/\mu L$	13.61 g/dL
4	Reference	$6.20 \times 10^3/\mu L$	$4.12 \times 10^6/\mu L$	$162 \times 10^3/\mu L$	13.20 g/dL
5	Control	$6.40 \times 10^3/\mu L$	$4.06 \times 10^6/\mu L$	$161 \times 10^3/\mu L$	12.90 g/dL

6.4 PRINCIPLE OF APPLICATION

This designed and developed Ag/CS/PVC nanocomposite blend biomaterial having overwhelming antimicrobial properties can effectively be used as self-sterilizing catheter material, especially for the cardiovascular system. Arjuna bark, one of the best cardioprotective agents, has been involved for the fabrication of silver nanoparticles reinforced into the nanobiomaterial. Antimicrobial assay shows significant bioactive characteristic of the developed nanobiomaterial.

Cross-linked blending of CS with PVC endows desired properties to the hybrid material. A cross-linking agent GA offered enhanced homogeneity and compatibility between the blended components. The interaction exhibited between the existing lone pair of electrons at the amino group of CS and the developed partial positive charge at the surface of silver nanoparticles due to electron drift, effectively stabilizes the silver nanoparticles and prevents them from agglomeration. But binding of silver nanoparticles with electrons of 'O' and 'N' atoms of hydroxyl and amine groups present in CS chains attributes lowering of swelling capability [50]. The blending with PVC produces additional crosslink's within the chain networks [51]. The higher crosslink's within the nanobiomaterial restricts the penetration of solvent for swelling. Nevertheless the enhanced CS content in the nanocomposite blend (2:1, CS: PVC) significantly increases the swelling capacity and imparts the required optimum hydrophilicity for bioactivity, restricting microbial growth and biofilm formation, while promoting fast healing through moderate absorption of wound exudates [52]. This is due to the presence of more hydrophilic groups in the networks which assist in improving the swelling characteristic of the composite nanobiomaterial [53].

6.5 CONCLUSION

Many synthetic and natural polymeric materials have limited applicability due to poor mechanical properties. Therefore, one of the chief objectives of this investigation was also to impart sufficient strength to the nanobiomaterial for superior biomedical applications. This nanobiomaterial also exhibited fairly good blood compatibility at specific levels through *in vitro* testing. Thus, this promising method of modification to combine CS with PVC and having nanobiotechnological herbal silver nanoparticles as antimicrobial agents incorporated into the copolymer matrix renders improved self-sterilization and finds potential application in biomedical engineering with the feasibility of this safe bioactive nanobiomaterial to be used as catheter biomaterial overcoming BAI.

REFERENCES

1. P. Dwivedi, D. Tiwary, P. K. Mishra and J. P. Chakraborty, Nano-Struct. Nano-Objects, Article ID 100485, 22 (2020) 1–7.
2. S. Dawadi, A. Katuwal, U. Gupta, R. Lamichhane, S. Thapa, G. Jaisi, D. P. Lamichhane, N. Bhattarai and Parajuli, J. Nanomater., Article ID 6687290, 2021 (2021) 1–23.

3. N. Jain, P. Jain, D. Rajput and U. K. Patil, Micro and Nano Syst. Lett., 9 (2021) 1–24.
4. P. Dwivedi, S. S. Narvi and R. P. Tewari, Ind. Crops Prod., 54 (2014) 22–31.
5. N. M. Alabdallah and M. M. Hasan, Saudi J. Biol. Sci., 28 (2021) 5631–5639.
6. P. Dwivedi, D. Tiwary, P. K. Mishra, S. S. Narvi and R. P. Tewari, Inorg. Chem. Commun., Article ID 108479, 126 (2021) 1–12.
7. C. K. Kokate, A. P. Purohit and S. B. Gokhale (eds.), Text Book of Pharmacognosy (41st ed), Nirali Prakashan, India (2008).
8. A. B. Choudhari, S. Nazim, P. V. Gomase, A. S. Khairnar, A. Shaikh and P. Choudhari, J. Pharm. Res., 4 (2011) 580.
9. T. R. Sobahi, M. S. I. Makki and M. Y. Abdelaal, J. Saudi Chem. Soc., 17 (2013) 245–250.
10. S. B. Bahrami, S. S. Kordestani, H. Mirzadeh and P. Mansoori, Iranian Polym. J., 12 (2003) 139–146.
11. S. Agnihotri, S. Mukherji and S. Mukherji, Appl. Nanosci., 2 (2012) 179–188.
12. H. S. Blair, J. Guthrie, T. Law and P. Turkington, J. Appl. Polym. Sci., 33 (1987) 641.
13. M. Kanke, H. Katayama, S. Tsuzuki and H. Kuramoto, Chem. Pharm. Bull., 37 (1989) 523.
14. M. Rinaudo, Prog. Polym. Sci., 31 (2006) 603.
15. P. Dwivedi, D. Tiwary, S. S. Narvi, R. P. Tewari and K. P. Shukla, Lett. Appl. Nanobiosci., 9 (2020) 1485–1493.
16. V. Titov, D. Nikitin, I. Naumova, N. Losev, I. Lipatova, D. Kosterin, P. Pleskunov, R. Perekrestov, N. Sirotkin, A. Khlyustova, A. Agafonov and A. Choukourov, Materials, 13 (2020) 4821.
17. I. Ahamad, N. Aziz, A. Zaki and T. Fatma, J. Appl. Phycol., 33 (2021) 829–841.
18. N. G. Kandile and A. S. Nasr, Carbohydr. Polym., 78 (2009) 753.
19. A. M. Rivera, K. W. Strauss, A. V. Zundert and E. Mortier, Acta. Anaesth. Belg., 56 (2005) 271–282.
20. E. M. Hetrick and M. H. Schoenfisch, Chem. Soc. Rev., 35 (2006) 780–789.
21. T. T. Trinh, P. A. Chan, O. Edwards, B. Hollenbeck, B. Huang, N. Burdick and J. A. Jefferson, Infect. Control Hosp. Epid., 32 (2011) 579.
22. R. M. Goda, A. M. El-Baz, E. M. Khalaf, N. K. Alharbi, T. A. Elkhooly and M. M. Shohayeb, Antibiotics, 11 (2022) 495, 1–12. https://doi.org/10.3390/antibiotics11040495
23. R. Varma, Medical Device Diagn. Ind., 19 (2007).
24. S. O. S. Bahaffi, M. Y. Abdelaal and E. A. Assirey, Intl. J. Polym. Mater., 55 (2006) 477.
25. M. Miya, R. Iwamoto and S. Mima, J. Polym. Sci. Polym. Phy. (22nd ed.), (1984) 1149.
26. T. M. Don, C. F. King and W. Y. Chiu, Polym. J., 34 (2002) 418.
27. M. Y. Abdelaal, E. A. Abdel-Razik, E. M. Abdel-Bary and I. M. Sherbiny, J. Appl. Polym. Sci., 103 (2007) 2864.
28. N. Minoura, T. Koyano, N. Koshizaki, H. Umehara, M. Nagura and K. Kobayashi, Mater. Sci. Eng., 6 (1998) 275.
29. W. Y. Chuang, T. H. Young, C. H. Yao and W. Y. Chiu, Biomaterials, 20 (1999) 1479.
30. M. Balouiri, M. Sadiki and S. K. Ibnsouda, J. Pharm. Anal., 6 (2016) 71–79.
31. G. Von White II, P. Kerscher, R. M. Brown, J. D. Morella, W. McAllister, D. Dean and C. L. Kitchens, J. Nanomater., Article ID 730746 (2012) 1–12.
32. J. Liu, M. Anand and C. B. Roberts, Langmuir, 22 (2006) 3964–3971.
33. S. K. Mehta, S. Chaudhary and M. Gradzielski, J. Colloid Interface Sci., 343 (2010) 447–453.

34. E. J. Mary and L. Inbathamizh, Asian J. Pharm. Clin. Res., 5 (2012) 159–162.
35. K. Shameli, M. B. Ahmad, A. Zamanian, P. Sangpour, P. Shabanzadeh, Y. Abdollahi and M. Zargar, Int. J. Nanomed., 7 (2012) 5603–5610.
36. S. M. Ghoreishi, M. Behpour and M. Khayatkashani, Physica E., 44 (2011) 97–104.
37. J. Huang, Q. Li, D. Sun, Y. Lu, Y. Su, X. Yang, et al., Nanotechnology, 18 (2007) 105104–105114.
38. K. Gopinath, S. Gowri, V. Karthika and A. Arumugam, J. Nanostruct. Chem., 4 (2014) 1–11.
39. K. Shameli, M. B. Ahmad, S. D. Jazayeri, P. Shabanzadeh, P. Sangpour, H. Jahangirian, et al. Chem. Cent. J.,6 (2012) 73–95.
40. L. Balau, G. Lisa, M. I. Popa, V. Tura and V. Melnig, Cent. Eur. J. Chem., 2 (2004) 638–647.
41. R. L. Williams, P. J. Doherty, D. G. Vince, G. J. Grashoff and D. F. Williams, Crit. Rev. Biocompat., 5 (1989) 221–223.
42. T. J. Berger, J. A. Spadaro, S. E. Chapin and R. O. Becher, Antimicrob. Agents Chemother., 9 (1976) 357–358.
43. R. M. Slawson, M. I. Vandyke, H. Lee and J. T. Trevors, Plasmid, 27 (1992) 72–77.
44. K. A. M. Amin and M. in het Panhuis, Polymers, 4 (2012) 590–599.
45. S. Azizi, M. B. Ahmad, M. Z. Hussein, N. A. Ibrahim and F. Namvar, Int. J. Nanomed., 9 (2014) 1909–1917.
46. P. Cazón and M. Vázquez, Environ. Chem. Lett., 18 (2020) 257–267.
47. S. Shrivastava, T. Bera, A. Roy, G. Singh, P. Ramchandrarao and D. Dash, Nanotechnology, 18 (2007) 225103–225111.
48. M. B. Kirschner, J. J. B. Edelman, S. C. H. Kao, M. P. Vallely, N. V. Zandwijk and G. Reid, Front. Genet., 4 (2013) 94–101.
49. L. A. Harker, B. D. Ratner and P. Didisheim, Cardiovasc. Pathol., 2, (Suppl 3) (1993).
50. A. Singh, M. Talat, D. Singh and O. N. Srivastava, J. Nanoparticle Res., 12 (2010) 1667–1675.
51. N. Ahmad, S. Sharma, M. K. Alam, V. N. Singh, S. F. Shamsi, B. R. Mehta and A. Fatma, Colloids Surf. B. Biointerfaces, 81 (2010) 81–86.
52. P. Dwivedi, S. S. Narvi and R. P. Tewari, Int. J. Sci. Res. Publ., 2 (2012) 1–5.
53. P. Dwivedi, S. S. Narvi and R. P. Tewari, Adv. Sci. Eng. Med., 6 (2014) 1–9.

7 Prospects of Safe Functional Nanomaterials in the Era of Virus Dominion

7.1 OVERVIEW

Undoubtedly, plant materials hold a unique ability for production of precise shape and controlled nanostructures, which needs to be up-surged, having the potentiality for myriads of applications. Presently the world is sailing in the horrifying scenario of the lethal pandemic of the coronavirus disease 2019 (COVID-19), with the causative novel corona virus, previously known as the 2019-novel coronavirus (2019-nCoV) while officially labeled as the severe acute respiratory syndrome coronavirus-2 (SARS-CoV-2). The global concern is deeply focusing virus transmission, pathogenesis and treatment, vaccine development, asymptomatic and presymptomatic virus shedding, while virus origin continues to be a research question. Prioritizing strategies to arrest the widespread community infection of this highly contagious deadly virus, which has seriously affected and seized the lives of human population umpteen worldwide, perceive to most crucially restrict the untoward. Hitherto, SARS-CoV-2 has crossed millions of humans tested positive confirmed cases and millions of deaths worldwide of COVID-19. Containment strategies and contamination minimizing measures adopting the highest possible advanced techniques with well-planned governance have fallen short, to hold control over the community spread at large, of this chiefly pneumonia like with variable manifestation of symptoms, extremely vulnerable contagious disease [1]. Coronaviruses (CoVs) belong to the family Coronaviridae, which are the enveloped viruses, having extraordinarily large ribonucleic acid (RNA) single-stranded genomes in the range of 26–32 kilobases length and 80–120-nm diameter. They are divided into four kinds: alpha-coronavirus (α-CoV), beta-coronavirus (β-CoV), delta-coronavirus (δ-CoV) and gamma-corona virus (γ-CoV) [1]. Reportedly seven types of CoVs affect humans; 229E, NL63, OC43 and HKU1 are associated with mild cold symptoms, whereas three of the kind β-CoV: Middle East respiratory syndrome coronavirus (MERS-CoV), Severe Acute Respiratory Syndrome Coronavirus 1 (SARS-CoV) and SARS-CoV-2 cause severe disease [2].

SARS-CoV-2 is the seventh member of CoVs known to infect human beings, targets angiotensin-converting enzyme 2 (ACE2), which is receptor expressed in the human lungs, gastrointestinal tract, cardiovascular and central nervous systems [3, 4]. The genome sequence of SARS-CoV-2 and SARS-CoV homology is around 79%, with interestingly highly similar receptor-binding domain (RBD) in their spike-protein, confirmed through several analyses. The RBD in the spike protein part is the most variable of the CoV genome, out of which six RBD amino acids have shown to bind ACE2 receptors critically and determine the host range for SARS-CoV viruses [5]. SARS-CoV-2 uses ACE2 as receptor, just like SARS-CoV, to recognize the corresponding receptor on the target cell by its S protein on the surface, thereafter enters into the cell, thereby causing the commencement of infection [1, 3]. Though, structural model assessment displays SARS-CoV-2 binding ACE2 with more than ten-fold higher affinity than SARS-CoV, much above the required threshold for virus infection [6]. Biochemical experiments and structural studies denote five among the six residues differ of SARS-CoV-2 from SARS-CoV that seemingly provide SARS-CoV-2 with RBD that binds with higher affinity to human ACE2 having high receptor homology [5]. While, computational analyses also predict the RBD sequence to be different from that of SARS-CoV for optimal receptor binding. The spike protein of SARS-CoV-2 having a functional, furin polybasic, cleavage site at the boundary of S1–S2, through the insertion of twelve nucleotides, which additionally leads to predicted acquisition of some three O-linked glycans available around the site, probably optimizes for binding to the ACE2 human receptor [5]. The detailed mechanism regarding the SARS-CoV-2 human transmission, interaction, infection and pathological organs damage, still remains the state-of-art for deep research and elaborate studies. Considering the higher binding affinity of SARS-CoV-2 with ACE2, further explains, the larger rapid human transmission aspect of SARS-CoV-2, than that by emerged SARS-CoV in March 2003 and MERS-CoV in 2012 [7–9].

Moreover, the unique features identified in SARS-CoV-2 genome, by the gene level systematic mutational analysis of its genomes, reveal idiosyncratic mutation, including spike surface glycoprotein (A930V (24351C>T)), occurring in some of its strains, also the immune epitopes predicted in the genomes, further complicates [10]. The heterogeneity of SARS-CoV-2 is intimidating with appalling COVID-19 condition, while conventional or advanced chemo-pharmaceuticals, Food and Drug Administration (FDA) approved medications, therapeutics and diagnostics, provide dismay [7]. Referring to earlier epidemiological studies, three conditions accountable for virus wide spread are the infection source, transmission route and susceptibility, with no exception for SARS-CoV-2 [1, 10]. Predominantly, four categorized treatment are recognized, Chinese medicine, antiviral western medicine, immune-enhancement therapy and convalescent plasma therapy with viral specific plasma globulin [1], albeit served for COVID-19 deserve larger clinical trials, determined safety consideration and applicable standards in clinical practice. SARS-CoV-2 is acutely virulent, keeping transmission dynamics through aerosol, droplets, close contact and even touch [1]. As the exact etiology, pathology and mechanism route remains unclear, specific drugs remain to be developed for this novel virus. Apparently it is prudent to interrupt the virus transmission route, in a bid for respite, to the widespread rapidly multiplying life-threatening SARS-CoV-2 and exponentially growing positive cases.

'Prevention is better than cure' as always, apprehending virus transmission dynamics, we focus on a preventive technique of contamination minimizing measure of maximized surface coating system with holy basil empowered self-sterilizing biocompatible antimicrobial silver/chitosan (Ag/CS) bionanocomposite, consisting intense antiviral aspects. The holy basil, indigenous to India, known as Tulsi, meaning the matchless/incomparable one, is regarded as the 'queen of herbs' and has botanical nomenclature of *Ocimum tenuiflorum* (*Ocimum sanctum* L.) [11]. Ayurveda reveres this holistic herbal plant as the 'mother medicine of nature' an 'elixir of life' and 'the herb for all reasons,' being possessed with perhaps equal levels of medicinal and spiritual healing properties [12]. The earliest reference of this whole plant (leaf, stem, flower, root and seed), constituted with safe medicinal therapeutic application values, is available in the quotes of Rig-Veda (3500–1600 BC) [13]. The other traditional medicine systems, such as Greek, Siddha, Roman, Unani and eventually modern medicine, do recognize its vast spectrum of potent pharmaceutical functions, which accounts: antiviral, antiprotozoal, antibacterial, antifungal/anticandidal, antimicrobial, anthelmintic, antiaflatoxigenic, antimetastatic, anticancer, radioprotective, antioxidant, antistress, immunomodulatory, antidiabetic, antidepressant, neuroprotective, hepatoprotective, cardioprotective, anticoagulant, analgesic, antipyretic, antiasthmatic, antibronchitis, antipneumonic, anticataract, cognition booster, renal damage recovery, wound healing, antiulcer, antiallergic, chemoprotective, dermatological purposes, etc. [13, 14]. Holy basil foliage down the ages has demonstrated immense pharmacological activities, imparting huge medical benefits, with extensive scientific evidences of its outstanding antiviral characteristics, due to the bioactive components enriched in particular. The essential extract of this foliage reportedly contains eugenol, ß-elemene, ß-caryophyllene and germacrene D as the major phytochemical constituents. Eugenol (1-hydroxy-2-methoxy-4-allylbenzene) is the biologically active compound responsible for mediation of pharmacological therapies and regulating physiological functions [11]. Ursolic acid and apigenin are the revealed active antiviral ingredients, conferring biological activity against deoxyribonucleic acid (DNA) viruses, such as herpes simplex viruses (HSV), adenoviruses (ADV) and hepatitis virus, as well as RNA viruses like coxsackievirus CVB1 and enterovirus 71 [15]. The much recommended abundant, economical, easily extractable, crude holy basil foliage extract has been investigated to be nontoxic as per methodical toxicity studies [16] and effective against paramyxo and orthomyxo viral pathogens importantly affecting humans [17]. The virucidal efficacy of the terpenoid and polyphenolic compounds, in the crude extract content, is also documented to resist influenza H9N2 virus infection [18]. The endowed inhibitory compounds have shown contradiction to several other viruses such as Vaccinia virus, Newcastle Disease virus, Infectious Bursal Disease virus, etc. [18]. Further corroborates, seven photophilic compounds studied from this basil namely, tulsinol A, B, C, D, E, F, G, along with dihydroeugenol B, denoting binding affinity with SARS-CoV receptors, meanwhile evident simulation confirms inhibition factor of holy basil extract in the replication of SAR-CoV [19]. The recent phytochemical evaluation and bioactive constituent identification have highlighted the relevance of the holy basil foliage extract against SARS-CoV-2, comprehending prospective key role to confiscate COVID-19 condition [20].

Based on enormous relevant experimental and published data, it has also been established that the inherent behavior of silver nanoparticles (AgNPs) is a useful potential tool against a range of viruses. Primarily, nipah virus (NiV), human immunodeficiency virus (HIV-1), H1N1 influenza A virus, HSV, hepatitis B virus, poliovirus type-1, respiratory syncytial virus (RSV), monkey pox virus, etc., have been dwindled by AgNPs [21–25]. The antiviral activity of Ag/CS bionanocomposite coating system shall be intensified, due to the coherent action of basil bioactive inhibitory compounds [25] that are persistently adhering to the silver (Ag) nanostructure surface, along with the other biomolecules, performing as the capping agents, for basil foliage extract powered formation of AgNPs [26, 27]. Chitosan (CS) biopolymer, thoroughly alone, doesn't exhibit any antiviral activity, instead provides a superb matrix phased base to the dispersed phase, facilitating proper reinforcement of the AgNPs, simultaneously reducing risk diffusion concerns of the AgNPs onto the environment [23, 28, 29]. The relationship existing between AgNP and its antiviral activity, determined by investigations suggest, an inversely proportional relation to its size, due to spatial hindrance, whereas directly proportional to its concentration, because of the manner in which the individual virus particles (virions) and nanostructures interact [23]. Albeit, the complex mechanism of action not vividly investigated particularly, usually the antiviral action of AgNPs against various virus types was demonstrated by blocking plasma membrane binding to virus-host cell, because of preferential direct binding of AgNPs to glycoprotein knobs at the bare sulfur-bearing residue sites present on the viral envelope, this inhibits virion penetration into the host cell [21–25]. AgNPs characterized as virucidal agents, also has been theorized to damage the viral genome, causing dismissal of viral replication [22]. AgNPs, even diminish host receptor and dependent associated co-receptor virion conformational changes, along with attachment and merging leading to pathogenesis, via viral interplay in cell-associated as well as cell-free states [28]. The slight electron drift in the valence band of AgNPs develops partial positive charge at the surface, to which the lone electron pair of the sulfhydryl group covering the viral envelop preferably interact [29, 30].

A cohesive strategy is presented here, as a convenient shield, in an effort to confront the virus transmission dynamics, via the means of maximized coating of exposed contact surfaces with self-sterilizing Ag/CS bionanocomposite material. Based on proven facts, collected evidences and exclusive research reports, our hypothetical proposal of self-sterilizing coating appears to be a prospective venture in the strive to control the thrive of SARS-CoV-2. Coating of surrounding contact areas and object surfaces having texture of metallic, ceramic, polymeric, wooden, etc., with Ag/CS bionanocomposite, will not alter the internal physico-chemical properties and original functionalities of objects. Rather cast outer layer of morphological modification in an amendment to render self-sterilizing antimicrobial nature externally, with an experimentally characterized biocompatible Ag/CS bionanocomposite material. The feasible coating system [28, 29], with this material of the holy basil empowered Ag/CS bionanocomposite, will unambiguously resist the virus SARS-CoV-2 [29, 30] to thrive on object surfaces, being microbe incompatible, reducing touch transmission through surface contamination. Green, cost-efficient and facile protocol of synthesis for Ag/CS bionanocomposite, aided with the beneficiaries of the holy basil foliage, has been followed to facilitate coating over maximum possible exposed object surface parts for the purpose. Applicable for publicly handled door handles, railings, seats, switch buttons, table tops, touch screens, key boards, etc., toward containment of the viral infection, in addition to substitute tedious and exorbitant sanitization routines and processes. The Ag/CS bionanocomposite nanomaterial was characterized through various techniques. Synthesis of Ag/CS bionanocomposite nanomaterial using the rhizome phytomass of *Curcuma longa* (turmeric) for the same purpose with similar aspect is noteworthy and referred [30].

7.1.1 SYNTHESIS – PHYTOMASS CONVERSION AND VALUE-ADDED FABRICATION OF FUNCTIONAL NANOMATERIALS

7.1.1.1 Materials

Silver nitrate ($AgNO_3$) of analytical grade, acetic acid glacial (extra pure), nutrient agar and nutrient broth were purchased from Thomas Baker (Chemical) Pvt. Ltd., India. Chitosan (CS, degree of deacetylation: 79%, molecular mass: 500,000 g/mol) was procured from Sea Foods(Cochin), India. Microbial strains *Pseudomonas aeruginosa* (Gram negative), and *Staphylococcus aureus* (Gram positive) were obtained from the culture bank of Microbiology Department, Sam Higginbottom Institute of Agriculture, Technology and Sciences, Allahabad, India. Foliage of the perennial herb *Ocimum tenuiflorum* (holy basil/Tulsi) was collected freshly from the surrounding. Deionized water was used for preparation of solutions.

7.1.1.2 Synthesis of Ag/CS Bionanocomposite Coating Material

Holy basil foliage, fresh, washed and air dried, weighing 20 g, was cut into fine pieces and boiled in 100 mL of sterile deionized water using a 500-mL Erlenmeyer flask and heating mantle, for 15 min at 100°C. The crude extract was filtered using funnel and filter paper (Whatman No. 41), thereafter 5 mL of the filtered holy basil extract (HE), mauve in color, was added to 100 mL of 1 mM colorless $AgNO_3$ solution. The reduction of silver cations (Ag^+) to silver atoms (Ag^0) and formation of AgNPs were demonstrated by very rapid color transformation from colorless to greenish brown, within 15 min, and was monitored by measuring the UV-visible spectrum of the reaction mixture (RM) having ($AgNO_3$ soluton + HE). After 48 h of complete stabilization at room temperature (RT), and when no further color transformation was observed, RM was centrifuged at 12,000 rpm for 15 min and the obtained residue (AgNP pellet) was re-dispersed into deionized water and re-centrifuged. This procedure was repeated to isolate AgNPs from loosely bound plant proteins and other unnecessary compounds present [13, 14]. The remnant pellet was dispersed in 15 mL of CS solution matrix (2% [w/v] in 1% [v/v] acetic acid), at specific ratio (1:5) and sonicated the dispersion for 15 min [30]. Finally, Ag/CS bionanocomposite was prepared to be used for coating of surfaces. The bionanocomposite material was synthesized following the green chemistry principles, avoiding harsh reagents that persistently adhere to the nanostructures and are hazardous to be handled as well as to be harnessed.

7.1.1.3 Coating of Object Surfaces

Coating of object surfaces [15, 16] was verified by solvent casting technique, i.e. pouring the Ag/CS bionanocomposite suspension onto stainless steel plates, rods and glass slabs, allowing to air dry, for preliminary studies which conferred favorable desired results. While, advanced coating techniques may be required for scale-up and probable to cast intended results. Ag/CS bionanocomposite thin film was obtained separately for several characterizations, by casting the suspension over glass slab, drying at RT and scratching out thereafter.

7.1.1.4 UV-Visible Spectrometric Study

UV-Visible spectrophotometer (Shimadzu UV-2450) was used to obtain the UV/Visible spectra ranging from 200 to 800 nm and govern the HE generated AgNPs and fabricated Ag/CS bionanocomposite coating material. The reduction of Ag^+ to Ag^0 was cautiously monitored through spectral data record from aliquots of RM. Spectra of only HE, $AgNO_3$ solution, Ag/CS bionanocomposite suspension and CS alone were also taken separately. Deionized water was used as blank.

7.1.1.5 HR TEM and SAED Observations with EDX Analysis

High Resolution Transmission electron microscopy (HR TEM) was performed on the HE generated AgNPs by placing drops of the RM suspension on carbon-coated copper grids and evaporating the solvent. Shape and size of the nanoparticles were investigated with the observations using (HR TEM TECNAI 20 G^2) instrument operated at an accelerating voltage of 200 kV and attached to an energy dispersive X-ray (EDX) analysis. EDX analysis identified the chemical composition of the nanoparticles. Selected area electron diffraction (SAED) spots of the AgNPs were also displayed.

7.1.1.6 Elucidation of Physico-Chemical Properties through FTIR, SEM and XRD

Fourier transformed-infrared spectroscopy (FTIR) using (Varian 3100) recorded between 400 and 4000 cm^{-1} enabled detail chemical composition elucidation of the AgNPs and Ag/CS bionanocomposite coating material. Vivid functional group identification was also facilitated for the determination of plant organic compounds and other biomolecules acting as stabilizing moiety, capping ligands and reducing agents. SEM analysis for morphological feature evaluation through (JEOL JXA 8100), while XRD pattern through (XRD, Philips, Xpert, Cu Kα) at a scanning speed of 3°/min for physical and structural characteristics of the Ag/CS bionanocomposite coating film were carried out.

7.1.1.7 Antimicrobial Assay

The HE aided Ag/CS bionanocomposite coating material was assayed for biological activity against *P. aeruginosa* (Gram negative), and *S. aureus* (Gram positive). Disk diffusion protocol was adopted to determine the standard zone of inhibition (ZOI). Ag/CS bionanocomposite coating film was cut into disk shape having 5-mm diameter and placed on different cultured agar plates. Antimicrobial test was also done against medical grade stainless steel plate (30 × 12 mm) piece (CSP), coated with the Ag/CS bionanocomposite. For positive and negative control, 400 mg of ≥98.0% Sparfloxacin powder (C+) having 11-mm disk diameter and distilled water (C–) in 7-mm diameter bored well were used, respectively. Nutrient agar was used as the culture media and inoculation was with ~900 μL of microbial organism containing broth. These plates containing the microbes and Ag/CS bionanocomposite coating film were incubated at 37°C for 48 h. Thereafter, plates were assessed for ZOI, which appears as a clear region around the disk. The ZOI diameters were measured using a meter ruler.

7.1.1.8 Cytotoxicity Test and Biocompatibility Assessment

In vitro cytotoxicity test was carried out following 3-(4,5-dimethylthiazol-2-yl)-2,5-diphenyltetrazolium bromide (MTT) assay protocol against the J774A.1 murine macrophage cell line, as a parameter for the evaluation of the biocompatibility extent, of the synthesized Ag/CS bionanocomposite coating material. Ag/CS bionanocomposite was placed as the test sample, while only macrophage cells as positive control and RPMI media as negative control were considered. Cells were placed at a density of 2.5×10^4 cells/well in a 96-well micro plate. Thereafter, the cells with varying concentrations were incubated in triplicate, for 72 h at 37°C, 5% CO_2 whereas untreated cells served as the control for the cytotoxicity test using the colorimetric MTT reduction assessment. Calculation of the 50% cytotoxicity concentration (CC_{50}) was from the graph of the optical density (OD) drawn against concentration, considering the OD of control well as 100% survival.

7.1.2 CHARACTERIZATION

7.1.2.1 UV-Visible Spectral Analysis

The UV-visible absorption spectrum recorded from the very rapidly HE generated AgNP suspension of RM, after 48 h of complete stabilization with no further color transformation, is shown in Figure 7.1. A surface plasmon resonance (SPR) band absorption peak λ_{max} appeared near 450 nm, which is characteristic of AgNPs between 420 and 480 nm [27, 28]. No absorption peak was observed around 220 nm in this spectrum, confirming the absence of unconsumed $AgNO_3$. The absorption spectrum of the corresponding HE did not exhibit such similar noticeable peak but have risen without any maxima or minima. The absorption spectrum of

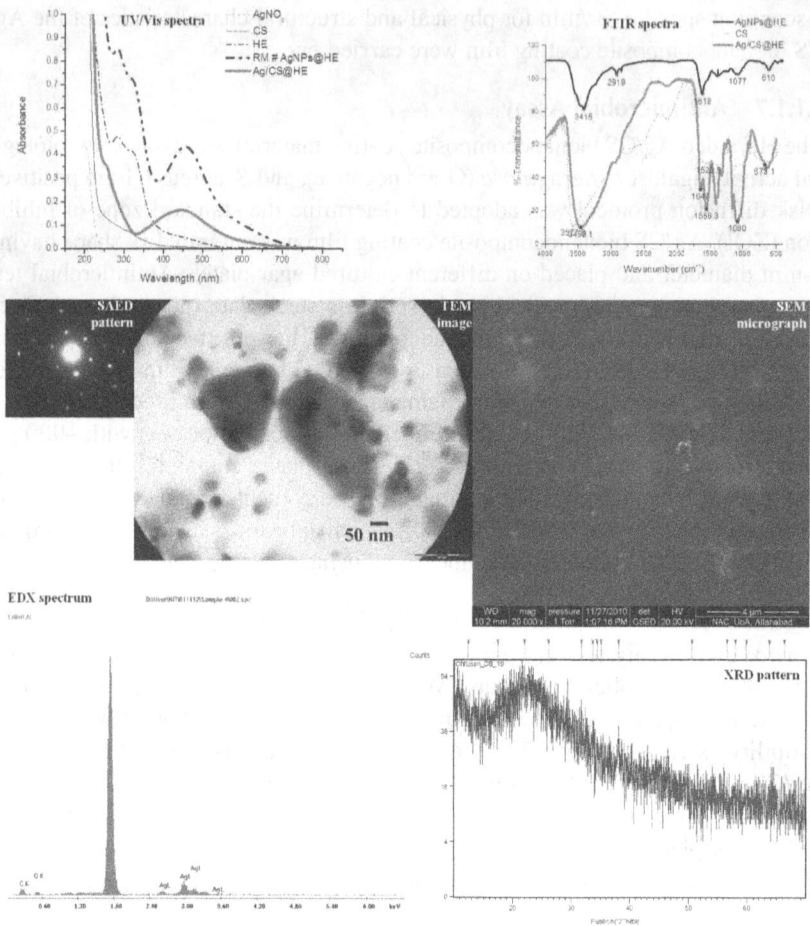

FIGURE 7.1   UV-Visible spectra of AgNO$_3$ solution, HE, RM # AgNP@HE suspension, CS solution and Ag/CS bionanocomposite suspension; FTIR spectra (%Transmittance/ Wavenumber cm^{-1}) of HE generated AgNP powder, 2% CS and Ag/CS bionanocomposite, films; SAED pattern and HR TEM image of HE generated AgNPs, SEM micrograph of the HE aided EDX spectrum of the AgNPs and XRD pattern of the Ag/CS bionanocomposite; respectively.

aqueous AgNO$_3$ only solution featured a rise around 220 nm. According to the generalized theory, only a single SPR band peak is expected for spherical nanoparticles, whereas anisotropic nanostructures or aggregates of spherical nanoparticles may have two or more SPR bands [27]. Spectral analysis of the synthesized Ag/ CS bionanocomposite coating material, also exhibited single absorbance peak similar to its counterpart of the AgNP aqueous suspension. Absorption spectrum of CS alone was transparent in the UV-visible region and did not have any maxima or minima.

7.1.2.2   Qualitative Assessment by HR TEM, SAED and EDX

The morphology of the HE generated AgNPs was observed by HR TEM and SAED images. The ring-like diffraction and approximately circular pattern of SAED bright spots reflected that the AgNPs were crystalline in nature (Figure 7.1). The HR TEM image, visualized in Figure 7.1, revealed that the AgNPs were not uniform in size, somewhat irregular in shape and there was presence of triangular plate like, miscellaneous structures but with chiefly spherical shapes in the nanorange. Indication of interparticle interactions was very less, which occurs due to the peripheral complexation of capped biomolecules, illustrating no agglomeration, denoting stability of the distinguished HE generated AgNPs and their prolonged use. Ag in the nanoparticles was evident from chemical composition identification by EDX analysis discussed in Table 7.1 and the EDX spectrum depicted in Figure 7.1. The presence of elements other than Ag identified by EDX is due to possible organic capping agents attached to the nanoparticle surface and interference of ions [27], during generation of the nanoparticles through bio-reaction by the varied phytochemicals of HE. The HE biomolecules, such as reducing sugars, terpenoid and polyphenolic bioactive compounds, responsible for the reduction of Ag^+ to Ag^0 and generation of AgNPs, further performed as capping ligands and stabilized the nanoparticles.

7.1.2.3   Study of Physico-Chemical Parameters
through FTIR, SEM and XRD

The active chemical species of HE intervening the AgNP generation are elucidated through FTIR spectroscopy; FTIR spectra illustrated in Figure 7.1 also consists of the spectrum exhibited by the AgNPs@HE. The large –OH and –CH stretches (3415 and 2919 cm^{-1}) observed were characteristic of sugar molecules, persistently being adhered to the AgNPs. The peak (1618 cm^{-1}) can be predominantly ascribed to the overlapping stretching vibrations of the –C=C– and –C=O characters, derived from the aromatic rings or polyphenolic groups. The peak at (1077 cm^{-1}) and (616 cm^{-1}) may be attributed to a range of stretching and bending vibrations belonging to amino acid functional groups. FTIR study confirmed

TABLE 7.1
Element Identification and Quantification
through EDX, of 'HE' Generated Nanoparticles

Element		
Carbon	Weight%	25.7
CK	Atomic%	42.9
Oxygen	Weight%	12.1
OK	Atomic%	33.5
Silver	Weight%	51.6
AgL	Atomic%	10.6

that protein residues, carbohydrates together with bioactive terpenoid and poly-phenolic molecules of HE acting as capping moieties and stabilizing agents were strongly adsorbed on the AgNP structure. The FTIR spectrum of Ag/CS bionano-composite in Figure 7.1 displayed intensified bending vibrations (1558–1080 cm^{-1}), indicating probable electrostatic interaction between AgNPs and amino groups of CS.

The HE empowered Ag/CS bionanocomposite in the SEM micrograph, dis-played in Figure 7.1, unveiled uniform dispersion of AgNPs with proper rein-forcement in the CS polymer matrix, imparting smooth nanomaterial surface, being unfavorable for micro-organism sustainability. Also, no aggregation was visualized, due to the interaction between the lone pair of electrons at the amine groups of CS and the partial positive charge developed by the slight electron drift at the outer surface of the AgNPs that effectively stabilizes the nanoparticles and prevents coagulation [29]. XRD pattern of the Ag/CS bionanocomposite, displayed in Figure 7.1, describes amorphous characteristic of the coating nano-material. Physico-chemical parameters of the Ag/CS bionanocomposite mate-rial unambiguously imply that coating of metallic, wooden, ceramic, polymeric, etc., textured object surfaces with this nanomaterial will render similar results, referred in our respective earlier studies [15, 16], and confer the presently con-cerned targeted purpose.

7.1.2.4   Antimicrobial/Biological Activity

The Ag/CS bionanocomposite coating material powered by HE was assayed for biological activity against *P. aeruginosa* (Gram negative) and *S. aureus* (Gram positive), which are staunch infection causing microbes. Detail results of the ZOI obtained through the adapted disk diffusion method are presented in Table 7.2. ZOI for the positive control, sparfloxacin (C+) was well recorded, while for the negative control, distilled water (C–), ZOI could not be formed against any of the strains. Although, the antimicrobial efficiency of Ag/CS bionanocompos-ite was well pronounced against both Gram-negative and Gram-positive bacte-ria, comparatively assessing the antimicrobial assay, ZOI was more prominent against *S. aureus*; therefore, used to further examine against Ag/CS bionano-composite coated object surface that achieved desired microbe inhibition. Gram-negative bacteria contains only a thin peptidoglycan layer of 2–3 nm between the

---

TABLE 7.2

Zone of Inhibition (ZOI) in (mm) against Selective Microbial Strains, *Staphylococcus aureus* and *Pseudomonas aeruginosa*; by Silver Nanoparticle Pellets Formed from Different Reaction Mixtures

Name of the Micro-Organism	C+	C–	CS	Ag/CS	CSP
Staphylococcus aureus	41.6	Nil	Nil	30	40
Pseudomonas aeruginosa	42.5	Nil	Nil	20	NA

---

cytoplasmic membrane and the outer membrane, whereas Gram positive ones lack the outer membrane but has a peptidoglycan layer of about 30-nm thickness [27]. The mechanisms of the stringent biological activity of AgNPs on microbes including a wide range of viruses and bacteria are hitherto not fully annotated. Albeit, the three most apprehended mechanisms of antimicrobial action are: uptake of free silver ions released from AgNPs, followed by interruption in ATP production and RNA or DNA replication, formation of reactive oxygen species (ROS), direct damage to cell wall, cell membranes, or outer coat and disruption. The antimicrobial activity of Ag has been known since time immemorial even before Hippocrates had recognized Ag for prevention of diseases and clinicians have accepted it for over more than 100 years.

The advent of nanotechnology has bestowed more systematic investigation for mode of antimicrobial action because of AgNPs being more efficient and casting superior self-sterilizing properties, due to large surface area to volume ratio. Our present observation gives the perception that several key factors are responsible for the extraordinary microbial inhibition; among which the chemical composition of nanoparticles with high percentage of Ag content has an important part, while the size distribution of AgNPs definitely plays a crucial role. The virucidal and bactericidal activity also depends on the stability of nanoparticles in the medium, and smaller sized particle having larger surface area to volume ratio are thus more effective. The class of material to which the synthesized, HE powered Ag/CS bionanocomposite coating material, belongs is capable of rendering virucidal and bactericidal efficacy enabling prolonged of Ag^+ biocide in concentration of < 0.1 ppm, to which only a very few strains can be intrinsically resistant. Ag^+ anchors to and penetrates the outer wall of the microbial biological structure, thereby modulating biological signaling by dephosphorylating putative key peptide substrates on tyrosine residues, finally, causes lysis by rupturing the cytoplasmic or biological structure of bacteria or viruses. Moreover, in the present context the inhibition properties of Ag/CS bionanocomposite coating material is empowered with the synergistic advantage of adherent HE bioactive molecules that greatly extends the self-sterilizing antimicrobial action and forecasts its enormous incompatibility to a broad variety of microbes including SARS-CoV-2.

7.1.2.5 Biocompatibility Assessment

The synthesized HE empowered Ag/CS bionanocomposite coating material is considered to be reasonably biocompatible and safe to be used in regulated quantity on the basis of experimentally carried out cytotoxicity test. The CC_{50} value of the Ag/CS bionanocomposite was $0.040 \pm 0{:}006$ mg/mL, while that of the only CS matrix was 2.892 ± 0.199 mg/mL. The cells alive in the wells allow development of the purple color. The gradual decrease in the development of purple color denotes gradient increase in the death of live cells. Intensity of the purple color has positive indication regarding the number of live cells in the wells. Wells in 6–8 column range in the middle of the MTT plate show slight increase in fading of purple color indicating minuscule cell death due to the presence of Ag/CS bionanocomposite material. Diminished purple color development in the wells is indicative of cytotoxicity. In the present study, the cytotoxicity of Ag/CS bion-anocomposite was only marginally higher than the cytotoxicity of the CS matrix. Henceforth, on the basis of experimental observation, it is hereby hypothesized that the synthesized Ag/CS bionanocomposite coating material powered by HE, having biocompatible nature shall be non-toxic to humans, simultaneously rendering incompatible features to a diversity of microbes including SARS-CoV-2.

7.1.3 ANTIVIRAL HYPOTHESIS

Corroborating the multimodal therapeutic properties of the holy basil, recent investigations also emphasizes its role as a modulatory agent in the prevention and control of COVID-19 [19, 20]. Exploring the part of its bioactive compounds for synergism enhancement of the biocompatible self-sterilizing antimicrobial characteristic of Ag/CS bionanocomposite, here we postulate the hypothesis for this coating material in restraining the transmission dynamics of SARS-CoV-2. Broad spectrum antimicrobial activity against an array of animal and human pathogens has also promoted the assorted applications of the holy basil scaling from hand sanitizer, mouthwash to water purifier, etc., offering ancient wisdom for solutions to modern problems, and suggesting the same diversified span covered by Ag and AgNPs. Therefore, when mankind is baffled in the battle with SARS-CoV-2 infection, insight into the cohesive antiviral action of the duo, adhered HE bioactive molecules contributively to AgNPs, presents a strategy conducive to confiscate the escalated menace of COVID-19 pandemic. Findings, toward interaction of viruses with antiviral nanoparticles, incorporated with individually distinct surface chemistry, specify marked differences in inhibition results, which are outlined in terms of dependence upon the capping groups developing the nanoparticles. Synthetic capping polymers commonly selected, encapsulate the nanoparticle surface, diminishing their affinity and inhibitory effect, whereas biomolecules as reducing and stabilizing agents for nanoparticle generation, render their essential binding surfaces practically free, enabling them to interact more intensely. Rather, in our proposition, the prudently developed coating material has the AgNPs embodied with the congruent beneficiaries of bound HE biomolecules although in traces, contributory to Ag, designed for long-term impact.

7.2 PRINCIPLE OF APPLICATION

While, the ability of the virus to engage its human cellular receptor, enter into the cells and replicate, apprehending a complex process, further prospective studies on the Ag/CS bionanocomposite coating aspect, focusing SARS-CoV-2 shall cast much light into this context. During this corona era facilitating the proposed green, cost-efficient, biocompatible self-sterilizing Ag/CS bionanocomposite facile surface coating system, feasible over maximum exposed surrounding contact areas, objects and parts, such as community handled door handles, railings, seats, table-tops, switch buttons, touch screens, key boards, etc., shall grant giant relief. The same also implies for hospital premises, medical equipment and devices. The methodology provides a substitute to exorbitant and tedious, regular and repeated processes of disinfection and sanitization. The HE empowered Ag/CS bionano-composite coating, casting minimal altered outward exterior to materials, is incompatible to microbes and prospectively to SARS-CoV-2, hence conflict their thrive onto the imparted self-sterilizing smooth surface modification, minimiz-ing susceptible contamination and community transmission. Study centered on constraining the virus transmission dynamics through novel antimicrobial mate-rial becomes inevitable when implementation of fruitful vaccination seems to be a distant affair due to the rapid emergent variants of SARS-CoV-2 popping out of mutation [31–37].

7.3 FUTURE SCOPE – FOCUSING ANTIVIRAL PERSPECTIVE DURING THE VIRUS DOMINION

Following the words of Richard P. Feynman, 'there is plenty of room at the bot-tom,' it was also observed that, 'there is plenty of room in the global garden.' The wealth of plant resources with bright research perspectives is yet to be prudently and pragmatically explored in the nano era.

Plant biomass acts as a reservoir of cofactors, coenzymes and reducing agents functioning as the electron donors. Hypothetically, these electron donors provide electrons for the reduction of silver ions (Ag^+) to silver metal atoms (Ag^0) in the process of nanoparticle production. Possible plant metabolites responsible for bio-reduction, capping and stabilizing the nanoparticles are hypothesized; but simul-taneously the question arises for quantitative estimation of the phytochemicals in the aqueous extract, exactly locating the bio-reducer and capping biomolecule, which places an avenue for future research.

Though many assumptions and theories have been laid down, however, the actual mechanism of phytofabrication of AgNPs through extracellular biosynthe-sis is still unknown. Henceforth, the scope of a sound and solid undeniable postu-lation justifying the mechanism of this biogenic route convincingly and sensibly remains further open.

Insight involving other faced beads of *Elaeocarpus ganitrus* Roxb., the Rudraksha, can impart light bestowing impressive results of nanoparticle pro-duction, following herbal route and include different innovative applications in future. The Rudraksha beads (seeds) possibly contain high levels of plant bio-molecules are rich in reducing agents and biocatalyst. A detailed study on these standards and phytochemical screening of the *Rudraksha* beads on this aspect is lacking, which is a point of deep research.

Among the nanocomposites developed via phytofabrication, it can be observed that the silver/chitosan/poly vinyl chloride (Ag/CS/PVC) nanocomposite blend demonstrates overwhelming antimicrobial activity, which is the center of atten-tion. The nanocomposites, Ag/CS/PVC blend as well as chitosan-g-poly acryl-amide (Ag/CS-g-PAAm) hydrogel, need proper cytotoxicity testing that could not be carried out due to solid nature of the samples, which requires different set up not available in the nearby laboratories and having certain limitations.

While, the idea of coating biomedical implants with the synthesized Ag/CS bion-anocomposite can be tested through advanced techniques of coating, which will definitely throw better results. Figuring future prospects of the nanobiotechnolo-gized, novel, cost-effective and easily approachable green nanomaterials for pos-sible applications focusing biomedical and life sciences with a thought of giving something crucial to mankind in this book, concerns do arise for the viral threats. The knowledge regarding the health impact together with the *in vivo* biocompat-ibility factor and *in vivo* antiviral assays against various virion strains, of the synthesized nanocomposites is an essential scope of study before they can be established and standardized as safe bioactive nanobiomaterials having favorable benefit-to-risk ratio.

REFERENCES

1. L. S. Wang, Y. R. Wang, D. W. Ye and Q. Q. Liu, Int. J. Antimicrob. Agents, 55 (2020) 105948.
2. N. Zhu, D. Zhang, W. Wang, X. Li, B. Yang, J. Song, X. Zhao, B. Huang, W. Shi, R. Lu, P. Niu, F. Zhan, X. Ma, D. Wang, W. Xu, G. Wu, G. F. Gao,W. Tan and China Novel Coronavirus Investigating and Research Team, N. Engl. J. Med., 382 (2020) 727–733.
3. Y. Qiu, Y. B. Zhao, Q. Wang, J. Y. Li, Z. J. Zhou, C. H. Liao and X. Y. Ge, Microbes Infect., 22 (2020) 221–225.
4. M. Olsen, S. E. Cook, V. Huang, N. Pedersen and B. G. Murphy, Int. J. Antimicrob. Agents, 55 (2020) 105964.
5. K. G. Andersen, A. Rambaut, W. I. Lipkin, E. C. Holmes and R. F. Garry, Nat. Med., 26 (2020) 450–455.
6. D. Wrapp, N. Wang, K. S. Corbett, J. A. Goldsmith, C. L. Hsieh, O. Abiona, B. S. Graham and J. S. McLellan, Science (New York, NY), 367 (2020) 1260–1263.
7. A. M. Zaki, S. van Boheemen, T. M. Bestebroer, A. D. Osterhaus and R. A. Fouchier, N. Engl. J. Med., 367 (2012) 1814–1820.
8. C. A. Devaux, J. M. Rolain, P. Colson and D. Raoult, Int. J. Antimicrob. Agents, 55 (2020) 105938.
9. J. Zheng, Int. J. Bio. Sci., 16 (2020) 1678–1685.
10. C. C. Lai, T. P. Shih, W. C. Ko, H. J. Tang and P. R. Hsueh, Int. J. Antimicrob. Agents, 55 (2020) 105924.
11. N. Bano, A. Ahmed, M. Tanveer, G. M. Khan and M. T. Ansari, J. Bioequivalence Bioavailab., 9 (2017) 387–392.
12. M. M. Cohen, J. Ayurveda Integr. Med., 5 (2014) 251–259.
13. S. S. Pingalea, N. P. Firke and A. G. Markandeya, J. Pharm. Res., 5 (2012) 2215–2220.
14. L. Mousavi, R. Mohd Salleh and V. Murugaiyah, Int. J. Food Prop., 21 (2018) 2390–2399.
15. L. C. Chiang, L. T. Ng, P. W. Cheng, W. Chiang and C. C. Lin, Clin. Exp. Pharmacol. Physiol., 32 (2005) 811–816.
16. N. Jamshidi and M. M. Cohen, Evid.-Based Complementary Altern. Med., Article ID 9217567, 13 (2017).
17. U. Patil, Int. J. Ayurveda Pharma Res., 6 (2018) 17–21.
18. S. S. Ghoke, R. Sood, N. Kumar, A. K. Pateriya, S. Bhatia, A. Mishra, R. Dixit, V. K. Singh, D.N. Desai, D. D. Kulkarni, U. Dimri and V. P. Singh, BMC Complement. Altern. Med., 18 (2018) 174–184.

19. S. Khaerunnisa, H. Kurniawan, R. Awaluddin, S. Suhartati and S. Soetjipto, Preprints, Version1 (2020) 2020030226.
20. A. K. Srivastava, J. P. Chaurasia, R. Khan, C. Dhand and S. Verma, Med. Aromat. Plants (Los Angeles), 9 (2020) 359.
21. J. L. Elechiguerra, J. L. Burt, J. R. Morones, A. Camacho-Bragado, X. Gao, H. H. Lara and M. J. Yacaman, J. Nanobiotechnol., 3 (2005) 6.
22. S. Galdiero, A. Falanga, M. Vitiello, M. Cantisani, V. Marra and M. Galdiero, Molecules, 16 (2011) 8894–8918.
23. Y. Mori1, T. Ono, Y. Miyahira, V. Q. Nguyen, T. Matsui and M. Ishihara, Nanoscale Res. Lett., 8 (2013) 93.
24. R. G. Kerry, S. Malik, Y. T. Redda, S. Sahoo, J. K. Patra and S. Mahji, Nanomedicine: NBM, 18 (2019) 196–220.
25. X. X. Yang, C. M. Li and C. Z. Huang, Nanoscale, 8 (2016) 3040–3048.
26. P. Dwivedi, S. S. Narvi and R. P. Tewari, International Conference on Nanoscience, Technology and Societal Implications, NSTSI11 (2011).
27. P. Dwivedi, S. S. Narvi and R. P. Tewari, Ind. Crops Prod., 54 (2014) 22–31.
28. P. Dwivedi, S. S. Narvi and R. P. Tewari, Int. J. Biomed. Nanosci. Nanotechnol., 2 (2012) 187–206.
29. P. Dwivedi, S. S. Narvi and R. P. Tewari, Nano LIFE, 5 (2015) 1540006.
30. P. Dwivedi, D. Tiwary, S. S. Narvi, R.P. Tewari and K. P. Shukla, Lett. Appl. Nanobiosci., 9 (2020) 1485–1493.
31. W. T. Harvey, A. M. Carabelli, B. Jackson, R. K. Gupta, E. C. Thomson, E. M. Harrison, C. Ludden, R. Reeve and A. Rambaut, COVID-19 Genomics UK (COG-UK) Consortium. In: S. J. Peacock and D. L. Robertson (eds.), Nature Reviews Microbiology, 19 (2021) 409–424.
32. D. Chakraborty, A. Agrawal and S. Maiti, The Lancet, 397 (2021) 1346–1347.
33. A. J. Greaney, T. N. Starr, P. Gilchuk, S. J. Zost, E. Binshtein, A. N. Loes, S. K. Hilton, J. Huddleston, R. Eguia, K. H. D. Crawford, A. S. Dingens, R. S. Nargi, R. E. Sutton, N. Suryadevara, P. W. Rothlauf, Z. Liu, S. P. J. Whelan, R. H. Carnahan, J. E. Crowe Jr. and J. D. Bloom, Cell Host & Microbe, 29 (2021) 44–57.
34. C. J. Gordon, E. P. Tchesnokov, R. F. Schinazi and M. Götte, J. Biol. Chem., Accelerated Communication Editors' Pick, Article ID 100770, 297 (2021) 1–8.
35. A. S. Lauring and E. B. Hodcroft, JAMA, 325 (2021) 529–531.
36. S. W. Huang and S. F. Wang, Int. J. Mol. Sci., Article ID 3060, 22 (2021) 1–21.
37. R. N. Tasakis, G. Samaras, A. Jamison, M. Lee, A. Paulus, G. Whitehouse, L. Verkoczy, F. N. Papavasiliou and M. Diaz, PLoS ONE, 16 (2021) e0255169, 1–23.

8 Nanobiotechnological Biomass Conversion into Other Metallic Nanostructures

8.1 INTRODUCTION

Nanomaterials, the broad class of extraordinary materials, including arrays of nanoscale level being constituted of nanoparticles which are the particulate substances having (1–100) nm range of one of the dimensions, consist (0D, 1D, 2D or 3D) forms, varied structural shapes, span of distributed sizes and diversified chemical characteristics [1]. Based on their chemical characteristics, the nanoparticulate structures and henceforth the nanomaterials can be approximately categorized into classes of carbon-based including graphenes, carbon nanotubes and fullerenes; metallic; ceramic; polymeric; semiconductor; lipid-based; etc. [1]. Based on physical classification of their composed phases they are categorized as nanocomposites composed of more than one phase out of which one being the matrix while others as dispersed or reinforced, nanoalloys consisting of coexisting phases, nanocrystals, etc. [2, 3]. The reactivity, stability, toughness and superior behavior of the nanomaterials greatly depend on their unique characteristic features of nanoparticulate shape and size distribution, as well as structural configuration. The morphological parameters are extremely and intensively influencing on the extensive applications they hold nowadays from commercial to domestic encompassing medical imaging to diagnostics and therapeutics, cumulating environmental, catalysis, energy, etc. [4].

In context, especially metallic, bimetallic and their corresponding oxide nanoparticles, exhibit far reaching advantageousness in profound applications that science, engineering and technology could uphold for household and industry. These do possess exuberant morphologies, such as nanostars, nanoprisms, nanopolyhedrons, nanocages, nanoboxes, nanodumbbells, nanoshuttles, nanodendrites, nanofibers, nanowires, nanorods, nanotubes, nanocylinders, nanorings, nanoplates, nanotriangles including the commonly developed nanospheres, etc., responsible for imparting unique properties conferring superior application oriented activities [5–7]. In pace with the trend of outreaching nanotechnology [8–13], foremost in cost effectiveness, facile syntheses, efficiency and applicability are nanoscale oxides of alkaline earth metals, e.g. magnesium oxide (MgO)

DOI: 10.1201/9781003217343-8

and calcium oxide (CaO) nanomaterials having vital applications in varied sectors [14–16].

Nano MgO and nano CaO have experimentally proven antimicrobial properties [17, 18], they act as functional nanomaterials essential for tissue engineering [19, 20], enable bone regeneration, while have also been explored as pivotal drug delivery agents [21–23], occupying crucial space in biotechnology and biomedical domain. Mg and Ca are dietary essentials due to their required abundance in the *Homo sapiens* (human) body; therefore, MgO and CaO nanomaterials are much in food industry demand, also because of being antimicrobial in nature endorsing food storage and for the purpose of packaging [24]. They even function as antioxidants [25, 26] sufficing the requirement in minimized elemental quantities of the nanomaterials pertaining to their outstanding characteristic exaggeration of activities causative of high (surface area: volume) ratio [6]. Nano MgO and nano CaO are also capable of promoting plant growth factors as well as superiorly support numerous functions in agriculture [27, 28]. Apart from biomedical and life sciences, nano MgO and nano CaO enormously contribute in the domains of energy, environmental, pollution control, purposing heavy metal adsorption, water remediation, toxicity removal and waste water treatment, performing techniques of advanced water purification, etc. [29–33]. Nano MgO and nano CaO are in high demand for catalysis [34–38], majorly for processes of energy product generation, e.g. pyrolysis [39] and transesterification reaction, for biodiesel production [40].

The human nail plates formed in influence and in contact with phalangeal bone's osteogenic layer, have identified deposits of Mg as well as Ca [41, 42], because of the fact that Ca is a major elemental constituent in bone composition whereas Mg has functional importance in bone metabolism. This justifies our significant innovative nanobiotechnological engineering of magnificent MgO/CaO nanoalloy comprising explicit nanostructures of nanorods alongside nanospheres, from remnant pruned-off nail plates present at the fingers and toes of human beings. The safe bioactive nanomaterial of MgO/CaO nanoalloy, bioengineered from renewable vestigial biomass of human nail pruning, which is otherwise discarded during regular personal grooming process, is an excellent valorization and presentation of waste to value conversion [43, 44].

Also to be mentioned, palladium (Pd), platinum (Pt), titanium (Ti), zinc (Zn), iron (Fe), copper (Cu), gold (Au) along with silver (Ag) as well as other metals, metalloids and non-metals in their elemental, oxide as well as hybrid forms in the nanostructured scale have crucial performances in diverse fields including exclusively in nanomedicine [45, 46]. Thus, after extensive review [47–53] and analysis for scope of fruitful application, we have investigated a divergent development of reforming nanomaterial involving palladium nanoparticles (PdNPs) through an altered chemical approach of wet impregnation pathway. Ultrastable Pd/Y-zeolite reforming catalyst nanomaterial is fabricated with the aids of nontoxic polymeric reagents, 2-(3,4-epoxycyclohexyl) ethyltrimethoxysilane (EETMOS) and polyvinylpyrrolidone (PVP). EETMOS a sol–gel (ormosil) precursor [54–56], functions as the reducing moiety facilitating the reduction of palladium cation (Pd$^+$) to palladium atom (Pd^0) causing the PdNP origination through bottom-up route by nucleation of Pd^0 particles. The effective size control of PdNPs is regulated by PVP molecules playing the role of stabilizers in order to prevent coagulation. The consequent modified construction of the Pd nanomaterial as a bifunctional reforming catalyst [57] is discussed herein briefly.

8.2   EXPERIMENTAL SECTION

8.2.1   MATERIALS

8.2.1.1   Materials for Nanobiotechnological Conversion of Biomass into MgO/CaO Nanostructured Alloy as a Nanomaterial

Human finger and toe nail pruning biomass, i.e. human nail biomass (HNB) was collected via personal grooming source as precursor material of magnesium nitrate [Mg(NO$_3$)$_2$] as well as calcium carbonate [CaCO$_3$]. Chemicals such as hydrogen peroxide (H$_2$O$_2$) used for the purpose was of analytical reagents (AR) grade and purchased in high purity from Merck/Sigma Aldrich, which required no further purification. Deionized water was used for specified purposes.

8.2.1.2   Materials for Pd Nanostructured Material Fabrication

Ammonium Y-zeolite (NH$_4$-Y) as a catalyst support, polyvinylpyrrolidone (PVP), 2-(3,4-epoxycyclohexyl) ethyltrimethoxysilane (EETMOS) and potassium tetrachloropalladate (II) (K$_2$PdCl$_4$) were purchased from Merck/Sigma Aldrich.

8.2.2   SYNTHESIS

8.2.2.1   Protocol for Nanobiotechnological Conversion of Biomass into MgO/CaO Nanostructured Alloy as a Nanomaterial

Calcination procedure [58] involving thermal decomposition was followed in accordance for the nanobiotechnological conversion of HNB into MgO/CaO nanoalloy as a nanomaterial. Prior to the proceeding the obtained HNB was washed thoroughly with soap and water, thereafter by the addition of few drops of H$_2$O$_2$ into water in order to remove the adhered impurities. The cleansed HNB was kept inside a hot air oven at 90°C for ~1 h, in order to dry the material and remove any loosely bound water molecules. The thoroughly dried HNB was pulverized using a grinder with intermittent operation of 30 min. The pulverized HNB considered as the raw material was weighed and thereafter calcinated using a muffle furnace and subjected to heat treatment at a mild temperature of 700°C for a limited duration of 1 h to obtain the resultant final product. The resultant whitish powdery product was characterized to confirm the nanostructures and elemental composition as MgO/CaO nanoalloy. Prior to characterization, the determined percentage yield calculated by following the equation: yield (%) = product weight (g)/raw material weight (g) × 100, was 62% approximately.

8.2.2.2   Protocol for Pd Nanostructured Material Fabrication

PdNPs formed following chemical reduction method, using K_2PdCl_4 as precursor, EETMOS as reducing agent while PVP as capping agent, were impregnated on Y-zeolite support through wet impregnation method. Addition of PVP (10 µL) to a solution containing 3:1 (K_2PdCl_4:CH_3OH) having K_2PdCl_4 (10 µL, 10 mM), and mixed, thereafter EETMOS (10 µL, 2.5 M) was added into the mixture and a vertex cyclomixer was used for stirring. The reaction mixture was allowed to incubate further for ~15 min, at ~45°C. After dark brown colloidal suspension was observed, confirming the formation of PVP stabilized PdNPs, fabrication of ultrastable Pd/Y-zeolite reforming catalyst nanomaterial was proceeded.

Commercial grade NH_4-Y zeolite was put to heat treatment in the presence of air for 4 h, at 450°C, at a heating rate of 1°C min^{-1}, to obtain the Y-zeolite support. Finally, Pd/Y-zeolite reforming catalyst nanomaterial was fabricated by the facile wet impregnation method on Y-zeolite support with PdNP colloidal suspension. Known amount of Y-zeolite support for this purpose was added into specific quantity of colloidal suspension containing PdNPs (1:10, wt/v) to consist a 10% (wt/v) mixed suspension. The reaction mixture was incessantly stirred vigorously for a period of 48 h with the help of a magnetic stirrer to allow uniform dispersion of metallic nanoparticles over the Y-zeolite catalyst support. Thereafter using a REMI centrifugation machine, the mixture was centrifuged accurately for 5 min, followed by removal of the supernatant with a micropipette and the obtained resultant brownish pellet was dried in the presence of air in an oven for ~6 h duration keeping at a temperature of 60°C. The achieved resultant nanomaterial, with ~22% metallic nanostructure loading on catalyst support material, was designated as Pd/Y-zeolite reforming catalyst nanomaterial [57].

8.2.3   CHARACTERIZATION

8.2.3.1   Characterization of the MgO/CaO Nanostructured Alloy as a Nanomaterial

The morphological and physico-chemical characteristics of the HNB nanobiotechnologized MgO/CaO nanoalloy were investigated by an energy dispersive X-ray spectroscopy (EDX) analyzer equipped field emission scanning electron microscopy (FE-SEM) system (FBI Nova NanoSEM 450). The EDX data was also obtained from the instrument along with the FE-SEM micrograph. Elucidation was rather vivid of the nanoalloy's nanostructures, which were observed through the high resolution transmission electron microscopy (HR-TEM) image and measurements, carried out using (FEI Type: TECHNAI G^2 20 TWIN), which was operating at (210–220 V, 50–60 Hz) accelerating voltage. Preparation of the nanoalloy sample for observation through HR-TEM was done by suspending minute amount of MgO/CaO nanoalloy powder into methanol, thereafter performing sonication in ultrasonic water bath for 5 min proceeded by depositing 1 drop of this suspension onto a copper grid (300-mesh), which was further allowed to dry at room temperature (RT).

8.2.3.2 Characterization of the Pd Nanostructured Material and Activity Assessment as Catalyst Nanomaterial

HR-TEM image and measurements of the PdNPs were obtained from the (FEI Type: TECHNAI G^2 20 TWIN) instrument operating at (210–220 V, 50–60 Hz) accelerating voltage, which was also attached to an EDX analyzer and EDX data was received from the instrument. EDX was performed by focusing the electron beam at different specific regions of the prepared PdNP sample. The preparation of PdNP sample for observation through HR-TEM was done by addition of minute quantity of colloidal PdNPs into methanol, thereafter sonication carried out in ultrasonic water bath for 5 min proceeded by depositing 1 drop of this suspension onto a copper grid (300-mesh), allowed to dry at RT. HR-TEM image captured was studied for the Pd nanostructure and size distribution prior to the wet impregnation step. The morphological features of the fabricated Pd reforming catalyst nanomaterial were investigated through an EDX equipped FE-SEM system (FBI Nova Nano SEM 450).

Activity of the Pd reforming catalyst nanomaterial was assayed through the nitrogen adsorption/desorption experiments using 'surface area and porosity analyzer' instrument (ASAP 2020 Micromeritics) to determine the Brunauer–Emmett–Teller (BET) surface areas as well as the pore volumes. Textural characterization, apparent surface area (S_{BET}) calculation of the Pd catalyst nanomaterial was performed by the means of nitrogen adsorption–desorption isotherms applying the BET equation. Pore dimensions were measured adopting the Barrett–Joyner–Halenda (BJH) advanced method. Prior to proceeding for measurements, the Pd catalyst nanomaterial sample was thoroughly degassed at 170–250°C for ~4 h *in situ* in vacuum in order to remove the previously adsorbed impurities. Total pore volume (V_{DR}) values were obtained by the Dubinin–Radushkevich (DR) equation applied to the data of N_2 (V_{DR} (N_2)) adsorption at a temperature of −196°C. The t-plot method was put to determine the micropore area and external surface area values of the Pd catalyst nanomaterial.

8.3 RESULTS AND DISCUSSION

The result obtained through FE-SEM analysis reveal remarkable distinct rod-shaped nanostructures with no agglomeration and display narrow size distribution in the micrographs (Figure 8.1). Miscellaneous nanostructures with somewhat irregular shapes as well as nanospheres were also visualized when observed from some other location point of the nanoalloy sample. Qualitative and quantitative description of the nanoalloy regarding the elemental configuration of the nanostructures generated from HNB through the nanobiotechnology pathway were obtained from recorded standard EDX spectrum, Table 8.1 presents the elemental

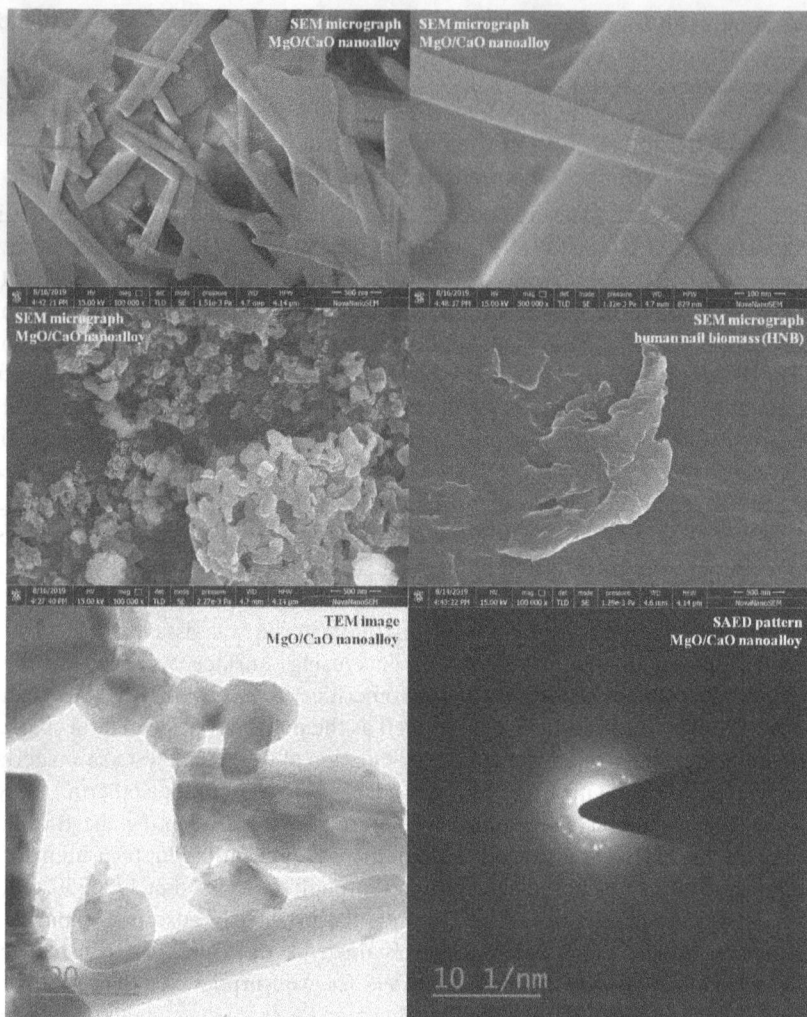

FIGURE 8.1   Illustration of the morphological and physical properties of the nanobio-technologized MgO/CaO nanoalloy through FE-SEM micrographs, HR-TEM images and SAED pattern.

composition of the nanoalloy, while Table 8.2 presents the elemental composition of HNB. EDX data confirms Mg and Ca existing in a homologous manner along with the presence of oxygen (O) in large quantity, which specifies the chemical composition of the nanostructures to be MgO/CaO, taking the form of nanostructured alloy, i.e. nanoalloy. The intensity of carbon (C) peak has been majorly contributed from the carbon tape requisite for carbon coating of the samples and placement in order to perform the scans. C contribution is also from the presence of C as residual organic ash in the content after calcination. Phosphorous (P) levels

TABLE 8.1
Elemental Composition of MgO/CaO Nanoalloy

Element	Weight (%)	Atomic (%)
Mg K	1.90	1.10
Ca K	8.91	3.20
C K	46.77	58.27
N K	0.01	0.01
P K	1.98	1.03
O K	40.43	36.39

indicate that phosphate present in the HNB undergoes decomposition and further functions as capping agent or stabilizer to the nanobiotechnologically obtained MgO/CaO nanoalloy so do the other organic biomolecules in order to prevent agglomeration of the nanostructures. It is evident that the capping biomolecules are in their respective oxides due to the high percentage of O. Stability, indicating prolonged use, as perceived from the magnificent morphology of the nanostructures worth notifying, consisting nanorods, nanospheres and miscellaneous shapes, displaying almost nil agglomeration, observed through the HR-TEM image (Figure 8.1). The diametric range of the nanorods was mostly in between (25–100) nm, whereas that of the nanospheres was largely in the range of (1–50) nm exhibiting size reduction. Selected area electron diffraction (SAED) pattern in Figure 8.1 shows prominent rings with many noticeable bright spots indicating crystalline nature. The bright spots usually arise from Bragg's reflection of the individual crystallite.

The developed Pd nanostructures as revealed from the HR-TEM scans were extremely minute in size in the form of nanodots, having narrow size distribution (d = 2–5 nm), with absolutely no agglomeration (Figure 8.2). The SAED pattern obtained, shown in Figure 8.2, displays diffuse rings attributing to the amorphous nature with scanty bright spots denoting crystallinity which came up from Bragg's reflection, but the ring made up of small spots specific for

TABLE 8.2
Elemental Composition of HNB

Element	Weight (%)	Atomic (%)
Mg K	0.06	0.04
Ca K	0.50	0.20
C K	65.17	71.02
N K	12.96	12.30
P K	1.10	0.46
O K	20.21	15.98

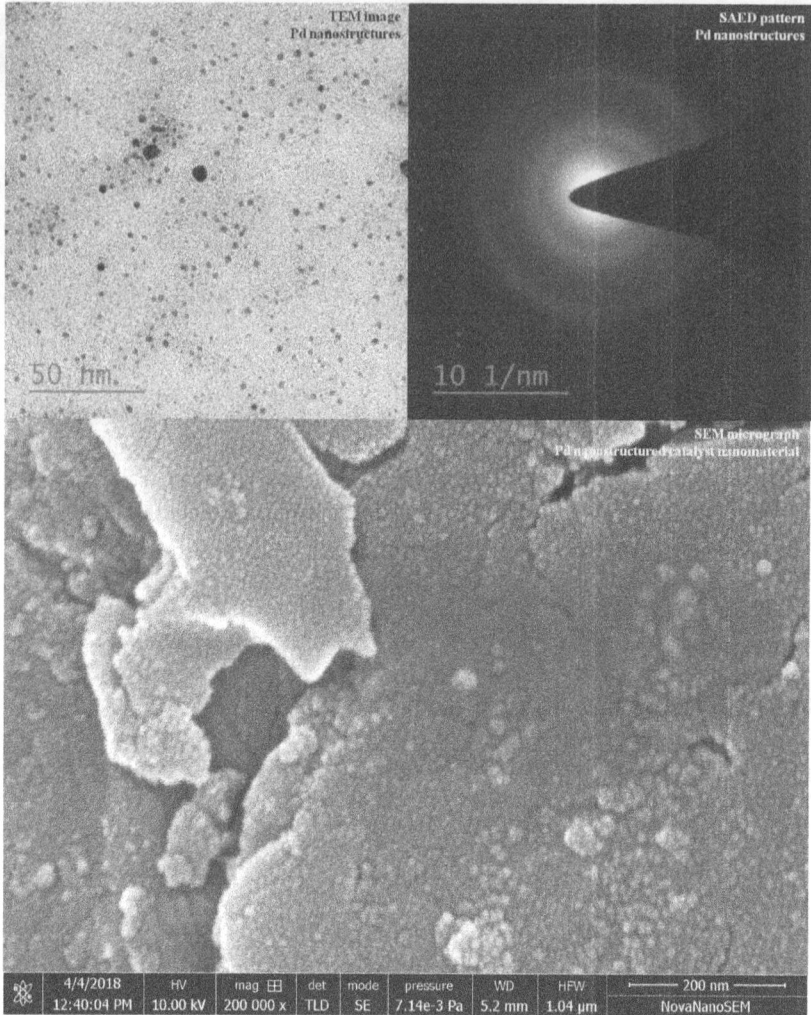

FIGURE 8.2 Illustration of the morphological and physical properties of the Pd nano-structures through the HR-TEM image and SAED pattern; morphological feature representation of the Pd nanostructured reforming catalyst nanomaterial designated as Pd/Y-zeolite through FE-SEM micrograph.

polynanocrystalline was absent. Qualitative and quantitative elemental configuration of the nanodots was obtained from the recorded standard EDX spectrum (Table 8.3). The presence of Pd homologously in the nanodots was confirmed by the corresponding peak. Other elemental belonging peaks also embody the nanostructures mainly as stabilizer moieties. FE-SEM analysis of the synthesized Pd nanostructured reforming catalyst nanomaterial presented in Figure 8.2 reveals uniform impregnation of PdNPs over the Y-zeolite catalyst support material.

TABLE 8.3
Elemental Configuration of Pd Nanostructures

Element	Weight (%)	Atomic (%)
Pd L	2.90	0.50
K K	0.10	0.05
C K	0.20	0.16
Si K	6.00	0.15
N K	11.70	13.82
O K	79.40	82.20
Cl K	0.29	0.52

EDX analysis confirmed Pd present over the surface of the catalyst nanomaterial sample (Table 8.4). A number of other elements, e.g. H, Na, Fe, Al, Si, O, S, etc., consisted in alumina-silicate complex structure of Y-zeolites also exhibited respective peaks. The weight% of the chief components were obtained by cautious elemental analysis having requisite degree of accuracy, with the spectrum recorded from a single-selected point that can vary and may not asserted to be the exact representative of the sample entirely throughout.

BET and BJH methods determined the in-depth textural characteristics of surface area and pore dimensions, of the Pd nanostructured reforming catalyst, as listed in Table 8.5. The catalyst nanomaterial exhibited a shape belonging to combination of type I (Langmuir) + IV isotherm classified for adsorption isotherms, feature of micro- and mesoporous structure [57]. Uptake of nitrogen was high at

TABLE 8.4
Elemental Configuration of Pd Nanostructured Reforming Catalyst Nanomaterial

Element	Weight (%)	Atomic (%)
Si K	33.06	23.99
Al K	12.45	9.40
C K	6.16	10.44
N K	6.07	8.82
O K	34.03	43.34
S K	3.98	2.87
Mg K	0.45	0.38
Ca K	0.58	0.29
Na K	1.80	1.60
K K	0.42	0.22
Fe K	0.54	0.22
Pd L	2.05	0.39

TABLE 8.5
Textural Properties of Pd Nanostructured Reforming Catalyst

Property	Pd/Y-Zeolite
BET surface area (m^2g^{-1})	457.4
t-Plot Micropore Area (m^2g^{-1})	434.7
t-Plot External Surface Area (m^2g^{-1})	22.7
BJH Adsorption average pore diameter (4V/A) (nm)	4.03
BJH Desorption average pore diameter (4V/A) (nm)	3.3
BJH Adsorption cumulative surface area of pores (m^2g^{-1})	16.8
BJH Desorption cumulative surface area of pores (m^2g^{-1})	25.5
BJH Adsorption cumulative volume of pores (cm^3g^{-1})	0.01
BJH Desorption cumulative volume of pores (cm^3g^{-1})	0.02

a very low relative pressure (P/P° < 0.1), which indicated microporous adsorption (type I) and a little capillary condensation occurred as the relative pressure was increased (type IV). Existence of a narrow hysteresis loop (H4 type), which is generally present in materials that have aggregates of uniform particles with slit shaped pores, attributed to mesoporosity.

8.4   CONCLUSION

Albeit in minuscule fraction, our research focused on the discarded remnant vestigial nail biomass conversion into value-added nanomaterial embracing nanobiotechnology, holds huge environmental impact reducing the waste load with high significance of reorganizing into valuable MgO/CaO nanostructures. The valuable nanostructures of MgO/CaO nanoalloy are bound to impart superior functionality regarding to bioactivity, advanced water treatment and remediation, catalyzes, dietary requirements, therapeutic and biomedical purposes, due to the dual advantage and composite properties of both MgO and CaO in combination than nano MgO or nano CaO working to benefit alone separately. The achieved MgO/CaO nanoalloy justifies the criteria of safe nanomaterials which can be defined as structures consisting one of its dimensions in the nanoscale of approximately (1–100) nm, having superior efficiency and possesses no toxic effects being bioengineered cost effectively in an eco-friendly manner. The nanorods and nanospheres are conspicuous and distinctly demonstrate negligible agglomeration with almost no requirement of any external reducing and stabilizing agent. This is possibly for the presence of a pool of requisite biomolecules in a particular nail biomass resource deployed as the required moieties in the nanobiotechnological fabrication process. The fabrication is of the kind of molecular self-assembly under physical and chemical phase deposition through the bottom-up synthesis approach. This is a scalable process nurturing molecular nano patterns in self-assembly with precise width and large pattern stretches. We have provided here a facile and versatile nanobiotechnological fabrication style allowing nanomaterial generation of complex multicomponent nanosystem, such as nanoalloy, exhibiting distinguished significant and magnificent nanostructures, composed with simultaneous deposition of MgO and CaO, consisting distinct characteristic activities. This is an ovel approach following the molecular self-assembly with plural materials, else hard to accomplish adopting other protocols, which is crucial for the design and development of diverse multifunctional nanosystem devices.

Nevertheless, the application of the synthesized Pd nanostructured reforming catalysts nanomaterial has potentiality to powerfully influence upon the distribution, selectivity, quality and yield of the products. The catalyst nanomaterial through optimization of the experimental conditions and reaction temperatures renders energy conservation and operating cost reduction. This research is based on the modification of reforming catalyst via wet impregnation of improvised Pd nanodots intended for constructing superior bifunctional nature to cast extraordinary characteristic features. Upgraded efficiency is speculated to be achieved with the resultant Pd catalyst nanomaterial in pilot-scale processes, prospected for industrial scale-up, to overcome the constraints of energy consumption and cost.

REFERENCES

1. I. Khan, K. Saeed and I. Khan, Arabian J. Chem., 12 (2019) 908.
2. P. K. Gautam, S. Shivalkar and S. Banerjee, J. Mol. Liq., 305 (2020) 112811.
3. S. Shivalkar, P. K. Gautam, S. Chaudhary, S. K. Samanta and A. K. Sahoo, J. Environ. Manag., 281 (2021) 111750.
4. W. Xu, Z. Li and Y. Yin, Small, 14 (2018) 1801083.
5. A. Gentile, F. Ruffino and M. G. Grimaldi, Nanomaterials, 6 (2016) 110.
6. A. G. Pershina, O. Y. Brikunova, A. M. Demin, O. B. Shevelev, I. A. Razumov, E. L. Zavjalov, D. Malkeyeva, E. Kiseleva, N. V. Krakhmal, S. V. Vtorushin, V. L. Yarnykh, V. V. Ivanov, R. I. Pleshko, V. P. Krasnov and L. M. Ogorodova, Nanomed., 23 (2020) 102086.
7. G. Costabile, R. Provenzano, A. Azzalin, V. C. Scoffone, L. R. Chiarelli, V. Rondelli, I. Grillo, T. Zinn, A. Lepioshkin, S. Savina, A. Miro, F. Quaglia, V. Makarov, T. Coenye, P. Brocca, G. Riccardi, S. Buroni and F. Ungaro, Nanomed., 23 (2020) 102113.
8. P. Dwivedi, S. S. Narvi and R. P. Tewari, Ind. Crops Prod., 54 (2014) 22.
9. J. A. Lee, M. K. Kim, H. M. Kim, J. K. Lee, J. Jeong, Y. R. Kim, J. M. Oh and S. J. Choi, Int. J. Nanomed., 10 (2015) 2273.
10. M. Navlani-García, I. Miguel-García, Á. Berenguer-Murcia, D. Lozano-Castelló, D. Cazorla-Amorósa and H. Yamashita, Catal. Sci. Technol., 6 (2016) 2623.
11. H. K. Kiranda, R. Mahmud, D. Abubakar and Z. A. Zakaria, Nanoscale Res. Lett., 13 (2018) 1.
12. D. Lombardo, M. A. Kiselev and M. T. Caccamo, J. Nanomater., Article ID 3702518, (2019) 1.
13. R. M. P. Kumar, A. Venkatesh and V. H. S. Moorthy, Nano-Struct. Nano-Objects, 21 (2020) 100406.
14. S. K. Moorthy, C. H. Ashok, K. V. Rao and C. Viswanathan, Mater. Today, 2 (2015) 4360.

15. A. Anantharaman, S. Ramalakshmi and M. George, Int. J. Eng. Res. Appl., 6 (2016) 27.
16. L. Habte, N. Shiferaw, D. Mulatu, T. Thenepalli, R. Chilakala and J. W. Ahn, Sustainability, 11 (2019) 3196.
17. P. Bhattacharya, S. Swain, L. Giri and S. Neogi, J. Mater. Chem. B., 7 (2019) 4141.
18. Y. Sato, M. Ishihara, S. Nakamura, K. Fukuda, T. Takayama, S. Hiruma, K. Murakami, M. Fujita and H. Yokoe, Molecules, 24 (2019) 3415.
19. E. A. Münchow, D. Pankajakshan, M. T. Albuquerque, K. Kamocki, E. Piva, R. L. Gregory and M. C. Bottino, Clin. Oral Investig., 20 (2016) 1921.
20. A. K. Singh, K. Pramanik and A. Biswas, Mater. Technol., 34 (2019) 818.
21. W. X. Tsai, P. A. Yu, S. S. Hwang, Y. W. Huang, F. Y. Hsu and Hsu, Pharmaceutics, 10 (2018) 179.
22. A. Alfaro, A. León, E. Guajardo-Correa, P. Reúquen, F. Torres, M. Mery, R. Sequra, P. A. Zapata and P. A. Orihuela, PLoS ONE, 14 (2019) e0214900.
23. S. M. Dizaj, S. Sharifi, E. Ahmadian, A. Eftekhari, K. Adibkia and F. Lotfipour, Expert Opin. Drug Deliv., 16 (2019) 331.
24. Y. He, S. Ingudam, S. Reed, A. Gehring, T. P. Strobaugh Jr. and P. Irwin, J. Nanobiotechnol., 14 (2016) 54.
25. R. Dobrucka, Iran J. Sci. Technol. Trans. Sci., 42 (2018) 547.
26. K. Kandiah, T. Jeevanantham and B. Ramasamy, Artif. Cells Nanomed. Biotechnol., 47 (2019) 862.
27. L. Cai, J. Chen, Z. Liu, H. Wang, H. Yang and W. Ding, Front. Microbiol., 9 (2018) 790.
28. S. Jha, R. Singh, A. Pandey, M. Bhardwaj, S. K. Tripathi, R. K. Mishra and A. Dikshit, Int. J. Res. Appl. Sci. Eng. Technol., 6 (2018) 460.
29. N. A. Oladoja, I. A. Ololade, S. E. Olaseni, V. P. S. P. O. Olatujoye, O. S. Jegede and A. O. Agunloye, Ind. Eng. Chem. Res., 51 (2012) 639.
30. Z. Camtakan, S. A. Erenturk and S. D. Yusan, Environ. Prog. Sustain. Energy, 31 (2012) 536.
31. M. Sadeghi and M.H. Husseini, J. Appl. Chem. Res., 7 (2013) 39.
32. B. Bharathiraja, M. Sutha, K. Sowndarya, M. Chandran, D. Yuvaraj and R. P. Kumar, Advances in Internal Combustion Engine Research, In: D. K. Srivastava et al. (eds.), Energy, Environment, and Sustainability, Springer Nature Singapore Pte Ltd., Singapore (2018) pp. 181.
33. N. Mazaheri, N. Naghsh, A. Karimi and H. Salavati, Iranian J. Biotech., 17 (2019) e1543.
34. N. Sutradhar, A. Sinhamahapatra, S. K. Pahari, P. Pal, H. C. Bajaj, I. Mukhopadhyay and A. B. Panda, J. Phys. Chem. C., 115 (2011) 25.
35. E. Mosaddegh and A. Hassankhani, Chinese J. Catal., 35 (2014) 351.
36. A. L. Gajengi and B. M. Bhanage, Adv. Powder Technol., 28 (2017) 1185.
37. S. M. Riyadh, K. D. Khalil and A. Aljuhani, Nanomaterials, 8 (2018) 928.
38. A. Zamani, A. P. Marjani and M. A. Mehmandar, Green Process. Synth., 8 (2019) 199.
39. A. K. Panda, R. K. Singh and D. K. Mishra, Renew. Sustain. Energy Rev., 14 (2010) 233.
40. R. A. Shokuhi, P. Mehdipour, A. Mirabi, A. Vaziri and E. Binaeian, J. Nanoanalysis, 4 (2017) 150.
41. B. Forsllnd, R. Wroblewski and B. A. Afzellus, J. Investig. Dermatol., 67 (1976) 273.
42. P. Saeedi, A. Shavandi and K. Meredith-Jones, J. Funct. Biomater., 9 (2018) 31.

43. P. Dwivedi, D. Tiwary, P. K. Mishra and J. P. Chakraborty, Nano-Struct. Nano-Objects, Article ID 100485, 22 (2020) 1–7.
44. P. Dwivedi, D. Tiwary, P. K. Mishra, S. S. Narvi and R. P. Tewari, Inorg. Chem. Commun., Article ID 108479,126 (2021) 1–12.
45. S. M. Saadati and S. M. Sadeghzadeh, Catal. Lett., 148 (2018) 1692–1702.
46. A. Ananth, V. Keerthika and M. R. Rajan, Curr. Sci., 116 (2019) 285–290.
47. M. S. Renzini, U. Sedran and L. B. Pierella, J. Anal. Appl. Pyrol., 86 (2009) 215–220.
48. N. Insura, J. A. Onwudili and P. T. Williams, Energy Fuels, 24 (2010) 4231–4240.
49. A. Lopez, I. Marco, B. M. Caballero, M. F. Laresgoiti, A. Adrados and A. Torres, Waste Manag., 31 (2011) 1973–1983.
50. W. Lutz, Adv. Mater. Sci. Eng., Article ID 724248:20, 2014.
51. P. Li, D. Li, H. Yang, X. Wang and H. Chen, Energy Fuels, 30 (2016) 3004–3013.
52. R. Miandad, M. A. Barakat, A. S. Aburiazaiza, M. Rehan and A. S. Nizami, Process Saf. Environ. Prot., 102 (2016) 822–838.
53. X. Lei, Y. Bi, W. Zhou, H. Chen and J. Hu, IOP Conf. Ser., 108 (2018) 022017.
54. D. Avnir, A. T. Coradin, B. O. Levc and J. Livageb, J. Mater. Chem., 16 (2006) 1013–1030.
55. G. Lakshminarayana and M. Nogami, J. Phys. Chem. C., 113 (2009) 14540–14540.
56. Y. Han, Q. Yu, J. Xu, J. Liu, Y. Yin and B. Li, Polym. Compos., 38 (2017) 657–662.
57. D. Dwivedi, P. K. Tiwary, J. P. Mishra and Chakraborty, Adv. Sci., Eng. Med., 12 (2020) 548–555.
58. Z. X. Tang, D. Claveau, R. Corcuff, K. Belkacemi and J. Arul, Matter. Lett., 62 (2008) 2096.

9 Environmental Fate of Safe Green Bioactive Nanobiomaterials

Life Cycle Assessment

9.1 CONCLUSION

Thus, there was an attempt to bring forth novel cost effective and easily approachable green nanomaterials for possible applications, especially in the important field of biomedical engineering, along with diversified purposes, with a thought of giving something crucial to mankind. The vitality of this work lies in the information it provides to the biomedical and nanotechnology based industries, for the design, development and outcome of value-added functional products, by utilizing the wealth of worldwide ecological resources [1]; eliminating obnoxious and toxic reagents which persistently adhere to the surface of the nanostructures and are rendered hazardous to be handled as well as to be applied, while following the (1, 3, 4, 5, 6, 7, 8, 10 & 12) principles of 'Green Chemistry.'

Furthermore, the prelude work mentioned in the preceding chapters of this book also concisely elucidates and imparts precious information regarding the investigated variations brought to the variables of silver nanoparticles through a variety of plant materials in the ecological diversity [1]. The dried fruits of *Elaeocarpus ganitrus* Roxb., i.e. the Rudraksha beads (endocarp with seeds enclosed within), synthesize silver nanoparticles gradually but with the capability to mediate many more number of syntheses, possessing the possibility for the preparation of several water extracts with the same beads, without any apparent disintegration and degradation, bestowing approximately similar results and high yield. It's this property of much delayed-degradability, unlike other perishable plant materials, that may make it the majesty of nanosilver generation in the nanoregime [2, 3]. There was also remarkable synthesis of silver nanoparticles from the foliage of *Prosopis spicigera*, i.e. the Shami plant, giving impressive antimicrobial properties [4].

Biosynthesis of silver/chitosan bionanocomposite via phytofabrication of silver nanoparticles with the foliage of *Pseudotsuga menziesii* (Christmas tree), to combat biomaterials associated infection (BAI), was applauded globally [5, 6]. It was felt noteworthy because the foliage of this plant are used in a huge scale for the Christmas decoration, they are removed after the celebration is over and otherwise wasted; we find here a fruitful purpose of their further consumption for

DOI: 10.1201/9781003217343-9

another good cause. Nanobiotechnological solution as a facile, cost-effective and efficient strategy to control the community spread of virulent contagions in the virus dominion has been postulated, hypothesizing *Ocimum tenuiflorum* (holy basil/Tulsi) foliage phytomass empowered bioactive nanomaterial surface coating system. *Curcuma longa* (turmeric) rhizome phytomass powered similar silver/chitosan bionanocomposite safe bioactive nanomaterial has been referred to for the related function exemplifying the scenario of coronavirus disease 2019 (COVID-19) pandemic [7]. Multifunctional MgO/CaO nanoalloy comprising amazing distinct nanorods and nanospheres, consisting safe bioactive characteristics, nanobiotechnologized through green pathway from human finger and toe nail pruning biomass (HNB), was magnificent and significant [8, 9].

9.1.1 NANOBIOTECHNOLOGICAL ROUTE AND ENVIRONMENTAL FATE

Nanobiotechnology and the development of nanobiomaterials including products consisting nanomaterials, popular as 'nanoproducts,' are rapidly developing in every field with myriads of innovative opportunities [10–15]. Howsoever, numerous uncertainties even do exist regarding the possible impacts of the nevertheless safe green nanobiomaterials synthesized through nanobiotechnological route on the human health and our environment. Therefore, utmost essential is of rather comprehensive and holistic assessment tools for 'life cycle assessment (LCA)' necessary to analyze, evaluate, comprehend, monitor and guard the environmental fate and health effects of the nanobiotechnologized products.

Due to great potential, smart and multifunctional performances of the nanomaterials for innumerable medical and industrial applications with prospective commercialization [16, 17], restrictions put forth by precautionary principles to avoid environmental and health impacts may lay constraints on the huge benefits gained even from nanobiotechnology. Measures implying protection of the environment from possible adverse effects caused by the nanomaterials along with identifying the societal effects to counterbalance the improvements from the nanoproducts themselves are the need of the time. Needless to be discussed, such as applications of nanobiotechnology as well as bionanotechnology in cancer treatment, varied areas of medicine, nanomaterials for water remediation, superior efficient energy systems, etc., not only encourages the society but also the individuals to accept the potential risks as the benefits posed by the nanoproducts outweighs.

Outline of the environmental (atmosphere, terrestrial and aquatic) pathways of nanomaterials developed employing nanobiotechnological route (NBNMs):

i. NBNMs released (intentional and unintentional) in the environment undergo different processes; combine with atmospheric constitutes to settle on terrestrial or aquatic planes
ii. NBNMs' atmospheric processes include
  a. Reduction
  b. Oxidation
  c. Heterocoagulation

    d. Diffusion
    e. Wet and dry deposition on terrestrial interfaces
  iii. NBNMs are bioavailable to plants from dissolution and sorption to roots
  iv. NBNMs' gravitational settling and adsorption on mineral surfaces leads to
    a. heteroaggregation within mineral layers
    b. dissolution within soil particles
  v. NBNMs are undergone heteroaggregation and sedimentation in the aquatic ecosystem, including
    a. Dissolution
    b. Surface coating over it by natural organic matter through sorption
    c. Binding to biota

Outline of the transformations of NBNMs – environmental fate:

  i. Sources of NBNMs into the environment
    a. Unintentional – NBNM product consumption, accidental spills, vehicular exhaust, atmospheric deposition, various industrial processes, etc.
    b. Intentional – Landfill waste deposition, waste water discharge, application of NBNMs for environmental remediation, NBNM-based agrochemicals, etc.
  ii. Transformation in the environment is regulated by the intrinsic properties of the NBNMs, such as their
    a. Shape
    b. Size
    c. Coating moieties
    d. Surface charge (positive, negative or neutral)
  iii. Environmental conditions of
    a. pH
    b. Presence of cations/anions
    c. Ionic strength
    d. Light
    e. Temperature
    f. Organisms
    g. Natural organic matter
    h. NBNM characteristics
  iv. Transformation processes include (elaborated in Table 9.1)
    a. Physical
    b. Chemical
    c. Biological
  v. Physical transformation processes of NBNMs include
    a. Adsorption
    b. Deposition
    c. Aggregation
    d. Miniaturization

TABLE 9.1

Environmental Fate of NBNMs – Physico-Chemical and Biological Transformations

Physico-Chemical Transformation		Biological Transformation	
Regulating factor	Transformation process	Regulating factor	Transformation process
NBNM concentration	Agglomeration	Microorganisms	Biodegradation
Brownian movement			
Gravitational force			
–	Homoaggregation	Microbial enzymes	Bio-reduction
			Bio-oxidation
NBNM size	Heteroaggregation	Biological mediated	NBNM surface coating
Surface charge		reactions	degradation
Van der Waals force			Phosphate
			transformation
Natural colloid interaction	Sulfidation	–	–
Natural organic matter sorption over NBNM surface	Surface coating of NBNM	–	–
Photochemical reactions	Generation of reactive oxygen species (ROS)	–	–
–	Dissolution	–	–

vi. Chemical transformation processes of NBNMs include
   a. Dissolution
   b. Interaction with pollutants
   c. Redox reactions
   d. Free radical reactions
   e. Photochemical reactions, etc.
vii. Biological transformation processes of NBNMs include
   a. Biodegradation
   b. Biochemical reactions
   c. Biomodification

Physico-chemical nanostructure characteristics, such as shape, size, surface area, surface charge/zeta potential, colloidal stability as well as core-shell composition, including environmental conditions of pH, ionic strength, temperature, organic and inorganic colloids, natural organic matter, etc., are the most important parameters which regulate the environmental transformations of nano-pollutants [18] from NBNMs. Therefore, organisms encounter in the environment, multiple transformed nanostructures rather than the engineered pristine NBNMs, because of the NBNMs' interaction with a range of environmental components [19]. Thus, the metrics and behavior of the transformed NBNMs is of utmost importance to realize their bioavailability, mode of toxicity and environmental fate.

9.1.2 Life Cycle Assessment (LCA) of Safe Green Bioactive Nanobiomaterials

In the present scenario, knowledge of the vivid exposure routes and the environmental risk impacts of NBNMs are limited. Additionally, the potential environmental advantages of the safe green bioactive nanobiomaterials over conventional nanomaterials have not been fully investigated. Therefore, establishment of a complete understanding of the environmental concerns and human benefits with respect to the drawbacks of nanobiotechnology and bioactive nanobiomaterials in comparison with the conventional nanomaterials and products through their complete life cycles is necessitated. LCA is an essential tool for achieving this goal, which is well-established to identify as well as quantify the environmental impacts of manufactured products together with their processes through their life cycles [20]. LCA is a set of standardized protocols for estimation and analysis of the resources, manufacturing, application, usage and environmental impacts, attributed to the whole life cycle of the nanoproduct [21]. In other words it is a 'cradle to the grave' assessment, starting from extraction, acquisition and processing of raw materials, through energy consumption, material manufacturing and production, to application and usage, leading to disposal, final transformation and end-of-life environmental fate [22]. Environmental impacts may include detriments of pollution, stratospheric ozone depletion, climate change, toxicological stress on the ecological diversity including humans and health risks, apart from concerns of resource consumption, water utilization, etc. Principally, the LCA protocol assays four phases, as schematically presented in Table 9.2, comprising of a series of guidelines for all the four phases [23].

In our research studies, biomass enabled nanostructures are the building blocks of the constructed safe green bioactive NBNMs. Extensive usage of such in various fields and especially in the biomedical arena due to their novel biological and physico-chemical characteristic features, may also cause concerns regarding the possible releases unintentionally or intentionally into the environment during every stage of the product life cycle. Speculated optimized impacts of the bioactive NBNMs being safer than their conventional nanomaterial counterparts have been concisely tabulated in Table 9.2.

As proposed by James Hutchison, Director of 'Safer Nanomaterials and Nanomanufacturing Initiative (SNNI),' world's leading green nanotechnology effort, toward nanotechnology Environmental Health and Safety (EHS) research in three phases of evolving approach [24], we have emphasized on the third, ensuring green nanoscience approach to bioactive nanobiomaterials and process design eliminating hazards from the entire nanomaterial's life cycle. We have innovatively incorporated, optimization of by-products, promotion of maximized recycling, minimization of wastes, etc., following the (1, 3, 4, 5, 6, 7, 8, 10 & 12) green

TABLE 9.2

Comparative Assessment of LCA of Conventional Nanomaterials and Safe NBNMs

Life Cycle Assessment (LCA)		
Life Cycle Stages	**Environmental and Human Concerns**	
	Conventional Nanomaterials	**Safe NBNMs**
Processing	Intensity of high energy	Low energy intensity
	Consumption of high resources (including water and other solvents)	Less resource consumption (mostly utilization of natural renewable resources)
	Occupational exposure to hazards	Avoidance of occupational exposure to hazards (eco-friendly)
Manufacturing	Usually consumption of high energy	Energy efficient
	Usually involvement of complex techniques and time consuming	Facile techniques involving short duration and time saving
	High cost	Cost-effective
	Occupational exposure to hazards	Avoidance of occupational exposure to hazards (eco-friendly)
Application	Harsh and often toxic exposure to the environment and humans	Non-toxic exposure to the environment and humans – comparatively safer
End	Recycling difficult	Easy recycling
	Usually non-biodegradable	Usually biodegradable
	Toxic releases into the atmosphere, terrestrial and aquatic systems	Possibly imposing negligible toxicity
	High risk impacts and human health consequences	Low risk impacts

alternatives from the principles of green chemistry, as outlined by Paul T. Anastas and John C. Warner, while advocated by the US Environmental Protection Agency (EPA) even for nanomanufacturing processes [25]. The EPA had also coined the term 'green chemistry,' to specify the principles [26]:

1. Prevention of waste
2. Atom economy maximization
3. Syntheses less hazardous
4. Safer chemical design
5. Safer solvents and auxiliaries
6. Energy efficiency design
7. Renewable feedstock utilization
8. Reduction of derivatives
9. Catalysis (vs. stoichometric)
10. Degradation design

11. Prevention of pollution by real-time analysis
12. Prevention of accidents through inherently benign chemistry

Our nanobiotechnological approach provides an avenue for the production of value-added goods consisting more unique benefits, for demanding societal services, keeping a holistic view to environmental problems. It is a presentation of the way humans can utilize renewable natural resources to design and engineer nanoconstructs for the industrial systems having harmonious interaction with the environment. While, bioengineering bioactive nanobiomaterial construct, we have strived to endorse the definition of 'sustainable development,' stated according to the World Commission on Environment and Development (WCED) report of 1987, *Development that meets the needs of the present without compromising the ability of future generations to meet their own needs* [27].

REFERENCES

1. P. Dwivedi, S. S. Narvi and R. P. Tewari, Ind. Crops Prod., 54 (2014) 22–31.
2. P. Dwivedi, S. S. Narvi and R. P. Tewari, Int. J. Green Nanotechnol., 4 (2012) 248–261.
3. P. Dwivedi, S. S. Narvi and R. P. Tewari, Adv. Mater. Res., 585 (2012) 144–148.
4. P. Dwivedi, S. S. Narvi and R. P. Tewari, Int. J. Adv. Eng., Sci. Technol., 2 (2012) 236–243.
5. P. Dwivedi, S. S. Narvi and R. P. Tewari, Int. J. Biomed. Nanosci. Nanotechnol., 2 (2012) 187–206.
6. P. Dwivedi, S. S. Narvi and R. P. Tewari, Nano LIFE, 5 (2015) 1540006.
7. P. Dwivedi, D. Tiwary, S. S. Narvi, R. P. Tewari and K. P. Shukla, Lett. Appl. Nanobiosci., 9 (2020) 1485–1493.
8. P. Dwivedi, D. Tiwary, P. K. Mishra and J. P. Chakraborty, Nano-Struct. Nano-Objects, Article ID 100485, 22 (2020) 1–7.
9. P. Dwivedi, D. Tiwary, P. K. Mishra, S. S. Narvi and R. P. Tewari, Inorg. Chem. Commun., Article ID 108479, 126 (2021) 1–12.
10. P. Dwivedi, S. S. Narvi and R. P. Tewari, International Conference on Nanoscience, Technology and Societal Implications, NSTSI11 (2011).
11. P. Dwivedi, S. S. Narvi and R. P. Tewari, Int. J. Eng. Res. Appl., 2 (2012) 1490–1495.
12. P. Dwivedi, S. S. Narvi and R. P. Tewari, Int. J. Sci. Res. Publ., 2 (2012) 1–5.
13. P. Dwivedi, S. S. Narvi and R. P. Tewari, Adv. Sci. Eng. Med., 6 (2014) 1–9.
14. P. Dwivedi, P. K. Mishra, M. K. Mondal and N. Srivastava, Heliyon, Article ID e02198, 5 (2019) 1–15.
15. P. Dwivedi, D. Tiwary, P. K. Mishra and J. P. Chakraborty, Adv. Sci. Eng. Med., 12 (2020) 548–555.
16. S. Gottardo, A. Mech, J. Drbohlavov, A. Malyska, S. Bowadt, J. R. Sintes and H. Rauscher, NanoImpact, Article ID 100297, 21 (2021) 1–10.
17. G. Chugh, K. H. M. Siddique and Z. M. Solaiman, Sustainability, Article ID 1781, 13 (2021) 1–20.
18. Q. Abbasb, B. Yousafa, M. U. Amina, M. A. M. Alid, A. Munir, J. El-Naggarf, M. Rinklebeg and Naushad, Environ. Int., Article ID 105646, 138 (2020) 1–18.
19. K. Ikuma, A. W. Decho and B. L. T. Lau, Front. Microbiol., Article ID 591, 6 (2015) 1–6.
20. G. Barjoveanu, O. A. Pătrăuțanu, C. T. Eodosiu and I. Volf, Sci. Reports, Article ID 13632, 10 (2020) 1–12.

21. R. Dhingra, S. Naidu, G. Upreti and R. Sawhney, Sustainability, 2 (2010) 3323–3338.
22. ISO 14040:2006: Environmental management – Life cycle assessment – Principles and framework, International Organization for Standardization (ISO).
23. ISO 14044:2006: Environmental management – Life cycle assessment – Requirements and guidelines, International Organization for Standardization (ISO).
24. J. E. Hutchison, ACS Nano, 2 (2008) 395–402.
25. Nanotechnology White Paper, EPA 100/B-07/001 (February 2007) www.epa.gov/osa.
26. P. T. Anastas and J. C. Warner, Green Chemistry: Theory and Practice, Oxford University Press, New York, NY (1998).
27. World Commission on Environment and Development; Our Common Future (The Brundt land Report); Oxford University Press, Oxford, UK (1987).

Index

For Product Safety Concerns and Information please contact our EU
representative GPSR@taylorandfrancis.com
Taylor & Francis Verlag GmbH, Kaufingerstraße 24, 80331 München, Germany

www.ingramcontent.com/pod-product-compliance
Lightning Source LLC
Chambersburg PA
CBHW070717220326
41598CB00024BA/3205